Animal Models in Radiotracer Design

Animal Models in Radiotracer Design

Edited by
RICHARD M. LAMBRECHT
WILLIAM C. ECKELMAN

With 57 Figures

Springer-Verlag
New York Berlin Heidelberg Tokyo

Richard M. Lambrecht
Chemistry Department
Brookhaven National Laboratory
Upton, NY 11973, USA

William C. Eckelman
Section of Radiopharmaceutical Chemistry
George Washington University Medical Center
Washington, DC 20037, USA

Library of Congress Cataloging in Publication Data
Main entry under title:
Animal models in radiotracer design.
Includes bibliographical references and index.
1. Radioisotopes—Therapeutic use. 2. Radioisotopes in medical diagnosis. 3. Animal models in research.
I. Lambrecht, Richard M., 1943- II. Eckelman, William C. [DNLM: 1. Radioisotopes—Diagnostic use—Congresses. 2. Drug screening—Methods—Congresses. 3. Models, Biological—Congresses, WN 445 A598 1982]
RM849.A54 1983 615.8′424′0724 83-10553

Softcover reprint of the hardcover 1st edition 1983

Typeset by BiComp, Incorporated, York, Pennsylvania.

9 8 7 6 5 4 3 2 1

ISBN-13:978-1-4612-5598-7 e-ISBN-13:978-1-4612-5596-3
DOI: 10.1007/978-1-4612-5596-3

Foreword

The editors have asked me to write this foreword probably for two reasons: First, I have spent 35 years trying to develop radiolabeled compounds that would help diagnose and treat disease. I, and the people from my laboratory, have been much involved with developing $Na^{131}I$ for the diagnosis and treatment of hyperthyroidism and well-differentiated thyroid cancer. Second, I was a participant throughout the entire symposium upon which this book is based.

Our latest ^{131}I-labeled compound, [^{131}I] metaiodobenzylguanidine, took us 13 years to develop. It has succeeded in diagnosing and treating adrenergic tumors under circumstances where $Na^{131}I$ would have failed for the diagnosis and treatment of thyroid cancer. I therefore hope that more scientists will be attracted to this field.

This book is written by intelligent, experienced investigators who are knowledgeable. It is a must for anyone new in the field. For the experienced, I would like to remind you that I attended the entire symposium; the breadth and timeliness of the update has helped us to redirect some of our current efforts into more scientifically productive channels.

William H. Beierwaltes, M.D.
Ann Arbor, July 1982

Preface

Radiotracers with potential applications in medical research and radiopharmaceuticals intended for use in clinical medicine are evaluated extensively in animals prior to human use. The choice of the animal model plays a critical role in the eventual success or failure of the animal studies in predicting *in vivo* selectivity of a radiotracer for a physiologic process, the biodistribution, and the pharmacokinetics in man. This book is a first attempt at gathering information on the effect of choice of animal species on the development of radiopharmaceuticals and radiotracers for biomedical research. Our intent was to produce a resource guide for both newcomers and experts in the radiopharmaceutical sciences. Chemists, physicians, pharmacists, cellular biologists, veterinarians, and technologists will find this first comprehensive compilation to be both informative and timely.

The publication of this book is the result of two successful workshops sponsored by the Radiopharmaceutical Science Council of the Society of Nuclear Medicine held in conjunction with meetings of the Society in Phoenix in January and in Miami Beach in June, 1982. Recognized experts were asked to identify, review, and critically explore how existing animal models are being effectively used and to evaluate examples of the successful and unsuccessful research with the models described in the vast and rapidly expanding literature of radiotracer development. The outstanding reception of the material presented in the lectures led to this book. All manuscripts were reviewed by referees.

Acknowledgments

William C. Eckelman is supported at George Washington Medical Center by the Department of Radiology and the following grants from the National Institutes of Health: HL-19127, CA-18675, CA-28462; and the U.S. Department of Energy, Contract No. ASO 5 82 ER 60039.

Richard M. Lambrecht acknowledges the continuing scientific collaborations and professional advice of Alfred P. Wolf. He extends special thanks and appreciation to Heidi Sylvestre for her encouragement and for being a friend. We gratefully acknowledge Lois Caligiuri for secretarial assistance.

Research at Brookhaven National Laboratory in Radiopharmaceutical Chemistry and Nuclear Medicine is supported by the U.S. Department of Energy and its Office of Health and Environmental Research. Research is also supported in part by the National Institute of Neurological and Communicative Disorders and Stroke.

William C. Eckelman
Richard M. Lambrecht
Upton, December 1982

Contents

Contributors

William H. Beierwaltes, Department of Internal Medicine, Division of Nuclear Medicine, University of Michigan, Ann Arbor, MI 48109, USA

Duane C. Bloedow, Department of Pharmacy, University of Colorado, Denver, CO 80202, USA

Edward A. Carr, Jr., Department of Pharmacology and Therapeutics, State University of New York and The Buffalo Veteran's Administration Medical Center, Buffalo, NY 14214, USA

William C. Eckelman, Division of Nuclear Medicine, Radiopharmaceutical Chemistry Section, George Washington University Medical Center, Washington, DC 20037, USA

Alan R. Fritzberg, Department of Pharmacy, University of Utah, Salt Lake City, UT 84112, USA

Brian M. Gallagher, Division of Pharmacology and Animal Studies, New England Nuclear Corporation, North Billerica, MA 01845, USA

Richard M. Lambrecht, Chemistry Department, Brookhaven National Laboratory, Upton, NY 11973, USA

Carla J. Mathias, Division of Radiation Sciences, The Edward Mallinckrodt Institute of Radiology, Washington University, St. Louis, MO 63110, USA

John G. McAfee, Department of Nuclear Medicine and Radiological Sciences, State University of New York, Upstate Medical Center, Syracuse, NY 13210, USA

Aldo Rescigno, Department of Chemistry, Brookhaven National Laboratory, Upton, NY 11973 and Yale University School of Medicine, New Haven, CT 06510, USA

Gopal Subramanian, Department of Nuclear Medicine and Radiological Sciences, State University of New York, Upstate Medical Center, Syracuse, NY 13210, USA

Michael J. Welch, Division of Radiation Sciences, The Edward Mallinckrodt Institute of Radiology, Washington University, St. Louis, MO 63110, USA

Leonard I. Wiebe, Department of Bionucleonics and Pharmaceutical Sciences, University of Alberta, Edmonton, Canada T6G 2N8

CHAPTER 1

Animal Models in Biomedical Research and Radiotracer Design

RICHARD M. LAMBRECHT, WILLIAM C. ECKELMAN, and ALDO RESCIGNO

Radiopharmaceutical science and the design of radiotracers to probe biochemical and physiological processes in animal models and man are rapidly growing fields (Eckelman, 1982; Fowler and Wolf, 1982; Lambrecht and Morcos, 1982; Glenn, 1982). Tracers labeled with short-lived radionuclides (radiopharmaceuticals) are finding applications in nuclear medicine for the diagnosis and treatment of disorders and disease. But ethical, legal, and practical constraints limit experimentation in humans such as described by Cohen (1953). As a result, the animal model provides a scientifically valid and moral approach to the development of knowledge and contributes to animal and human welfare (NAS, 1981). It is not uncommon that the enthusiastic and well-intended researcher following the suggestion of a clinician has prepared a radiotracer in search of a disease, rather than having proceeded by analytical dissection of the scientific literature to arrive at a decision as to which radiotracer should be targeted against which disease or fundamental question in a carefully chosen animal model. "An animal model of a disease is only as useful as the questions we ask of it" (Reid, 1980). Regardless of the nature of the experiment, the experimentalist is required to incorporate the rational selection of a suitable animal model(s) as part of the experimental design.

Leader and Padgett (1980) and Jones (1980) have validated the historical justification of animals as models in the investigation and management of infectious human disease such as smallpox, tuberculosis, rabies, malaria, and influenza. Hevesy's book (1948) is recommended for a historical perspective of the application of radiotracers in biochemistry, animal physiology, and pathology.

An example that demonstrates the development of radiotracers and the transition from animal studies to human experimentation is the recent development of the deoxyglucose method (Sokoloff, 1977) for the *in vivo* measurement of local cerebral glucose utilization. The [^{14}C]DG autoradiographic studies with several species were terminal experiments that were a necessary prelude to nondestructive tracer kinetic and physiologic modeling experiments in nonhuman primates and man that were made possible by the advent of positron emission tomography (PET). See for example Lambrecht and Rescigno (1983). PET technology permits the quantitative reconstruction of the distribution of radioactivity within a given volume element (Ter-Pogossian, 1981; Ter-Pogossian et al., 1975; Phelps, 1981). PET is based on the coincident detection of annihilation radiation; therefore the tracer must be labeled with a radionuclide that decays by positron emission. ^{14}C decays by beta emission, and thereby requires that the animal be sacrificed in order that *in vivo* autoradiographic information can be obtained. Carbon-11 and fluorine-18 decay by positron emission with a half-life of 20.5m and 110m, respectively, and are appropriate radionuclides for incorporation into DG for use with PET.

The theory and procedure for the simultaneous measurement of the rates of glucose consumption in the various structural and functional components of the brain of the conscious and anesthetized albino rat were developed with ^{14}C-labeled 2-deoxy-D-glucose ([^{14}C]DG). The tracer monitors the exchange of glucose between plasma and brain and its phosphorylation by hexokinase in the tissues (Sokoloff et al., 1977). The method works because [^{14}C]deoxyglucose-6-phosphate is essentially trapped in the tissue during the course of the experiment. A kinetic model was developed (Sokoloff and Smith, 1983) that has subsequently been applied to the cat, monkey, and other species. Buchner and Buchner (1981) have discussed the extension of the deoxyglucose method for insects where electron microscopic resolution is required. The [^{14}C]DG method has been applied to understanding cognitive process (LePoncin-Lafitte, 1982) and various functional processes in animals (Sokoloff, 1983).

Ito, Wolf, and co-workers (1978) subsequently prepared 2-fluoro-2-deoxy-D-glucose labeled with fluorine-18 [2-^{18}F]DG and validated (Reivich et al., 1978; Phelps et al., 1979; Huang et al., 1980) the Sokoloff technique for *in vivo* autoradiography of man by utilization of PET. MacGregor et al. (1981) subsequently prepared [^{11}C]DG. The literature is exponentially expanding with PET investigations in man with [2-^{18}F]DG and [^{11}C]DG (Reivich and Alavi, 1983). Local cerebral responses to the normal unstimulated state (Mazziotta et al., 1981) and to visual, auditory, and other somatosensory stimuli are being studied in normal subjects (Alavi et al., 1982; Reivich & Alavi, 1983; Greenberg et al., 1981; Schwartz et al., 1983; Phelps et al., 1981; Mazziotta et al., 1982). Other investigators are addressing local glucose utilization in patients with specific diseases such as stroke (Kuhl et al., 1980), senile dementia (Ferris et al., 1980; Farkas et al., 1982), Alzhemier's disease (DeLeon et al., 1983), and various psychiatric disorders (Farkas et al., 1980; Farkas et al., 1981).

There is an absence of animal models for the human abiotrophies: Huntington's chorea, Parkinsonism, and amyotrophic lateral sclerosis (Vogel, 1980). Therefore certain radiotracer studies must be performed in man for the benefit of mankind.

Another developing example is the research of Crawford et al. (1982), who employed several small-animal models to evaluate ^{18}F-labeled-5-fluorouridine as a probe for measuring the proliferation of tissue *in vivo*. The fundamental studies in animals may lead to a predictive method in combination with PET such that therapy in man can be studied with a pulse-labeling technique.

Gill (1980) has classified the major types of animals for use in biomedical research. (1) Randomly bred animals for first-level chemical and tracer screening from colonies or wild populations. (2) Specifically structured outbred populations with a stabilized gene pool for first- and second-level screening. (3) Inbred strains of F1 hybrids for investigation of specific experimental questions and for comparing results over a long period of time, and genetic analysis. (4) Congenic strains for studying the effects of specific genes and their alleles against a common inbred background. For example, there are more than 400 well-studied hereditary errors in man. A high percentage of these are associated with cerebral dysfunction, and a defective gene has been attributed to each. Many have a model in lower animals (Vogel, 1980).

Animals have provided valuable models in biomedical research and the choice of the model should not necessarily be limited to mammals. Figure 1-1 is a guide to the frequency of use of animals in experimental medicine.

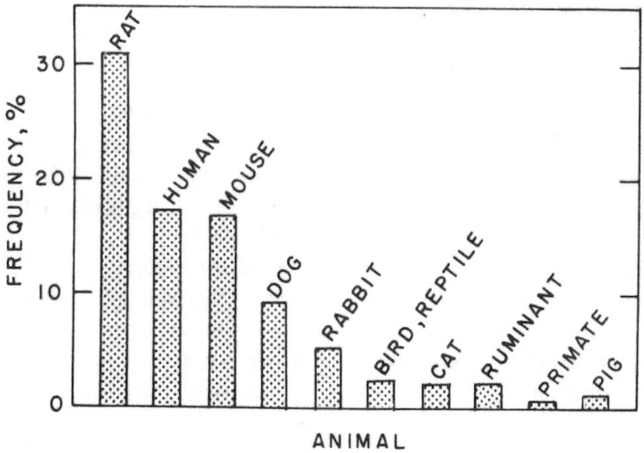

FIG. 1-1. Frequency of use of animals in experimental medicine as derived by Schmidt-Nielsen (1982) by a survey of the 4758 abstracts reporting the use of animals in the 1981 issues of *Federation Proceedings*.

For example, amphibians are the animals of choice for membrane studies. Table 1-1 lists 48 selected subjects in biomedical research with animal models that are provided with comprehensive reviews and references. It is an impossible goal to identify all the models and their proposed applications in biomedical research. Clearly, arrival at a decision as to the success or failure of an animal to model specific disease or biochemical process is an awesome chore requiring a task force of experts. The proceedings of the task force on models appropriate for studying diabetes mellitus and its complications is a notable example of the effort required to review the state of the art with respect to a disease for which a radiotracer can be designed to answer fundamental questions.

TABLE 1-1. Selected Books, Monographs, Reports and Reviews Devoted to Animal Models as Predictive Systems in Biomedical Research

Topic	*Reference*
Aging	NAS, 1981
Alcohol dependence	Pohorecky, 1981
Alcohol research	Ericksson, 1980
Aquatic animals as models	Bend and Weber, 1979
Behavioral effects of lead exposure	Bornschein et al., 1980
Blood–brain exchange	Fenstermacher, 1980
Cancer chemotherapy	Schabel, 1975
Cardiology	Tanzer, 1981; Harmison, 1973
Circadian rhythms	Berridge et al., 1979
Dental research	Navia, 1977
Dermatology, dermatopharmacology, dermatotoxicology	Maibach, 1975
Diabetes mellitus and its complications	Salans and Graham, 1982
Drug action	Sassenrath and Chapman, 1976
Environmental pollutants	NAS, 1979
Epilepsy	Purpura et al., 1972
Emphysema	Karlinsky and Snider, 1978
Endocrine system	Capen, 1980; Cahill, 1980
Eye	Albert, 1980; Aguirre, 1980
Fetal alcohol syndrome	Abel, 1982
Fetal medicine	Nathanielsz, 1980
Gastrointestinal system	Arias, 1980; Cheville, 1980
Hematology	Harmison, 1973
Herpes virus associated malignancy	Hinze, 1974
Graft-versus-leukemia	Okunewick, 1981
Hypoxia	Stefanovich, 1981
Immunity and disease	Gershwin and Cooper, 1978
Immunological processes	Hay, 1982; Smith, 1979
Invertebrates as models	Prosser, 1973
Lesser bushbaby (Galago) model	Haines, 1982

TABLE 1-1. (Continued)

Topic	*Reference*
Mammals as models	Gärtner et al., 1982
Medical research	Gill, 1980; Harmison, 1973; Jones et al., 1972–1980; Lindsey and Capan, 1976; Mitruka et al., 1976; NAS, 1969
Metabolic controls	Berdanier, 1975
Neurological disease	Rose, 1981
Neurology	Hanin and Usdin, 1977
Nervous system	Hadlow, 1980; Vogel, 1981
Obesity	Festing, 1979
Pig model	Dodds, 1982
Psychiatry	Hanin and Usdin, 1977; Ingle and Shein, 1980
Psychobiology	Serban and Klig, 1976
Psychopharmacology	Pieper, 1976
Reproduction, contraception, fertility	Alexander, 1979; Benirschke, 1980; Kumar, 1980; Serio and Martini, 1978
Respiratory system	Harmison, 1973; Reid, 1980; Slauson and Hahn, 1980
Spontaneous models	Andrews et al., 1979; Bustad et al., 1976
Surgery and transplantation	Harmison, 1973
Thrombosis and hemorrhagic disease	DHEW, 1976
Toxicological research	Bartosek et al., 1982
Tumor blood circulation	Peterson, 1979
Zucker-fatty rat model	Bray, 1977

This volume is the first to address animal models and their selection from the viewpoint of radiopharmaceutical science and nuclear medicine. "If we look carefully enough, we will eventually find an animal model for every disease" (René Dubos in Albert, 1980). Table 1-2 lists 190 specific diseases and biochemical processes and the animal models that have been documented as useful models of human disease and biochemistry. Table 1-2 is a detailed supplement to the current literature and not a composite of the information found in Table 1-1. Table 1-2 is necessarily incomplete because of the extensive literature on animal models. The intent is to provide the experimentalist involved in radiotracer design, radiopharmaceutical development, and nuclear medicine with a summary of information obtained in various areas of biomedical research so that new developments and rational fertilization may occur. The models for the specific topics discussed in the following chapters, for example, cancer and oncology (Wiebe, Chapter 4), are not intentionally duplicated in Table 1-2.

TABLE 1-2. Animal Models Proposed for the Investigation of Human Disease

Disease	*Animal Model*	*Reference*
Ocular melanoma	nu/nu Mice, Syrian hamster, rabbit, rat, kitten	Albert, 1980
Rheumatoid arthritis	Dog	Allen and Newton, 1975
Gastric ulcer	African rodent	Andre and Andre, 1981
Ulcerative colitis	Induced in guinea pigs with degraded carrageenan	Anver and Cohen, 1976
Liver cell carcinoma	Rainbow trout	Ashley, 1973
Retinitis pigmentosa	Rat, mice, dog	Aguirre, 1980
Adenocarcinoma	Aged rabbits	Baba and von Haam, 1972
G_{M1} gangliosidosis	Siamese cat	Baker and Lindsey, 1974
Glomerulonephritis Autoimmunity, autoantibody	Horse	Banks, 1979
Defects of iron transport	Mice	Bannerman, 1976
Valve implantation	Sheep	Barnhard et al., 1981
Chronic bronchitis	Swine	Baskerville, 1976
Hirschpung's disease, aganglionic or hypoganglionic megacolon	Piebald and spotted mouse strains	Bolande, 1975
Carcinoma of the ureter and urinary bladder	BN/BiRij rats	Boorman et al., 1977
Medullary carcinoma of thyroid	Rat	Boorman and Hollander, 1976
Cystic fibrosis	Oyster	Bowman et al., 1980
Obesity	Several models	Bray, 1977
Skeletal muscle glucose, glycogen metabolism	Rat	Brumback et al., 1980
Acrodermatitis enteropathica, zinc malabsorption	Cattle	Brummerstedt et al., 1977
Sensitivity to polychlorinated biphenyls	Gunn rats with hereditary hyperbilirubinemia	Calabrese, 1978
Prenervous functions of neurotransmitters	Early sea urchin embryo	Buznikov et al., 1982
Sensitivity to ozone	Mice with low levels of glucose-6-phosphate dehydrogenase	Calabrese, 1978
Diseases of the endocrine system	Rat, mice, Chinese hamster, dog, swine, sheep, goats, cattle	Capen, 1980

TABLE 1-2. (Continued)

Disease	*Animal Model*	*Reference*
Medullary thyroid carcinoma, multiple endocrine neoplasma, Sipples syndrome	Ultimobranchial thyroid neoplasm in bull	Capen and Black, 1974
Cushing's syndrome, hypercortisolism, Nelson's syndrome	Dog	Capen and Martin, 1975
Tardive dyskinesia	Rhesus monkey, guinea pig	Carlson, 1977
Oral Yersinia enterocolitica	Mice	Carter, 1975
Carcinoma of urinary bladder	Capuchin monkey	Cheever et al., 1976
Gonorrhea	Chimpanzee	Chandler and Kraus, 1976
Pneumocystis pneumonia	Immunosuppressed rat	Chandler et al., 1979
Diabetes mellitus—juvenile and maturity onset types	Mice infected with M variant of encephalomyocartis virus	Craighead, 1975
Diseases of the gastrointestinal system	Inbred lines of rodents, dog, cat	Cheville, 1980
Cardiopulmonary hypertension	Stumptail monkey	Chesney and Allen, 1973
Dysbaric osteonecrosis	C57BL/6J mice	Chryssanthou, 1981
Carcinoma of the urinary bladder	Rats induced by *N*-[4-(5-nitro-2-uryl)-2-thiazolyl]formamide	Cohen and Friedell, 1979
Motor neuron disease	Hereditary canine spinal muscular atrophy	Cork et al., 1980
Crigler–Najjar syndrome	Hereditary nonhemolytic unconjugated hyperbilirubinemia in Gunn rats	Cornelius and Arias, 1969
Acute lymphoblastic leukemia, aplastic anemia	Cat	Cotter, 1977
Idiopathic polyneuritis, Guillain–Barre syndrome	Coonhound paralysis	Cummings and Haas, 1972
Gold thioglucose obesity syndrome	Mice	Debons et al., 1977
Endodermal sinus tumor (ovaries, testes)	Rat	Damjanov, 1980

TABLE 1-2. (Continued)

Disease	*Animal Model*	*Reference*
Teratoma and teratocarcinoma	Mice	Damjanov and Solter, 1976
Megaloblastic anemia	Temperature-sensitive mutant mouse L cells	Dardick et al., 1978
Multiple sclerosis	Theiler's virus infection in mice	DalCanto and Lipton, 1977
Immune-mediated thrombocytopenia, autoimmune thrombocytopenia, idiopathic thrombocytopenic purpura	Dog, pig, horse	Dodds and Wilkins, 1977
G_{M1}-gangliosidosis, Type II	Cerebrospinal lipidosis of Friesian cattle	Donnelly and Sheahan, 1975
Experimental autoimmune myasthenia gravis	Rabbit	Eldefrawi, 1978
Glucose metabolism in eye	Rabbit, dogfish, sculpin	Edelhauser et al., 1980
Hepatic coma	Rat, dog, monkey	Fischer, 1977
Congenital hemolytic anemia	Alaskan malamute dog	Fletch and Pinkerton, 1973
Globoid cell leukodystrophy, Krabbe's disease	Dog	Fletcher and Kurtz, 1972
Congenital erythropoietic porphyia	Fox squirrel	Flyger and Levin, 1977
Adrenal injection, necrosis and hypercorticism	Golden hamster	Frenkel, 1977
Adenosis and clear cell carcinoma of vagina and cervix	Mice	Forsberg, 1976
Glaucoma	Beagle	Gelatt et al., 1981
Myocardial failure, muscular dystrophy	Cardiomyopathic Syrian hamster	Gertz, 1973
DiGeorge syndrome, congenital thymic hypoplasia	Congenitally athymic (nude) mouse	Gershwin, 1977
Angioimmunoblastic lymphadenopathy with dysproteinemia	New Zealand mice	Gindhart and Greenspan, 1980
Intraventricular hemorrhage	Beagle puppy	Goddard et al., 1980; Ment et al., 1982

TABLE 1-2. (Continued)

Disease	*Animal Model*	*Reference*
Hyperactivity	Rat	Goldberg and Silbergeld, 1977
Hashimoto's thyroiditis	Dog	Gosselin et al., 1978
Autosomal trisomy, developmental impairment, and fetal death	Mice	Gropp, 1975
Solar keratosis and squamous cell carcinoma	Beagle dog	Hargis and Thomassen, 1979
Pulmonary fibrosis	BALB/C mice	Haschek et al., 1981
Mucopolysaccharidosis	Siamese cat	Haskins et al., 1981
Effects of xenobiotics	Several species of fish	Hawkes, 1980
Hemolytic anemia of premature infants associated with vitamin E deficiency	Monkeys	Hayes, 1974
Ehlers–Danlos syndrome	Dogs and mink	Hegreberg, 1975
Muscular dystrophy	Mink	Hegreberg et al., 1976
Experimental carcinogensis	Rainbow trout	Hendricks et al., 1980
Persistent viral infections	Mink	Henson and Gorham, 1973
Pulmonary alveolar lipoproteinosis	Silica-induced pulmonary alveolar lipoproteinosis	Heppleston, 1975
Renal tubular carcinoma	Fischer 344 rats	Hinton et al., 1980
Infectious and neoplastic respiratory disease	Mice	Holt et al., 1978
Menkes' kinky hair syndrome	Mice	Holtzman, 1976
Carcinoma of the larynx	Susceptible inbred Syrian hamster	Homburger et al., 1979
Angiosarcoma	Cat	Hong et al., 1980
Melanoma	Swine	Hook et al., 1982
Burkitt's lymphoma, malignant lymphoma, and lymphocytic leukemia	Herpes virus—induced in squirrel monkey	Hunt and Melendez, 1972
Pemphigus vulgaris (skin/mucous membranes)	Dog	Hurvitz, 1980
Pituitary tumors	Rat	Ito, 1976

TABLE 1-2. (Continued)

Disease	*Animal Model*	*Reference*
Amebic dysentery	Germfree guinea pig	Jervis and Takeuchi, 1979
Genital tract infections	Baboon, chimpanzee, rhesus, grivet and Taiwan monkey, marmoset, pigtail macaque	Johnson and Taylor-Robinson, 1981
Mannosidosis of children, other inherited lysomal storage disorders	Pseudolipidosis of Angus cattle	Jolly, 1974
Central nervous infection	Eosinophilic meningoencephalitis in mice	John and Martinez, 1975
Hereditary muscular dystrophy	Chickens	Julian, 1973
Dietary-induced overeating	Rat	Kanarek and Hirsch, 1977
Acute lymphoblastic leukemia	Guinea pig	Kaplow and Nadel, 1979
Acute nonbacterial gastroenteritis	Dogs infected with coronavirus-I-71	Keenan et al., 1979
Hydronephrosis, agenesis of the renal medulla, arterial hypertension	Rat	Kentera et al., 1977
Diabetes mellitus	Inherited early-onset insulin-requiring diabetes in Keeshound dog	Kramer, 1981
Epilepsy	(*Papio papio*) baboon	Killam, 1976
Spina bifida, sacral dysgenesis, and myelocele	Manx cats	Kitchen et al., 1972
Tardive dyskinesia	Neuroleptic-induced, Sprague Dawley rats	Klawans et al., 1977
Dyskinesia	Dopamine agonist-induced guinea pig	Klawans et al., 1977
Struvite urolithiasis	Miniature Schnauzer dog	Klausner et al., 1981
Subacute sclerosing panencephalitis, multiple sclerosis	Distemper-associated demyelinating encephalomyctitis	Koestner, 1975
Hypertension, obesity, Type 4 hyperlipidemia, vascular disease	Rat	Koletsky, 1975

TABLE 1-2. (Continued)

Disease	*Animal Model*	*Reference*
G_{M2}-gangliosidoses	Dog	Karbe, 1973
Congenital hydrocephalus	Rhesus monkeys	Krous et al., 1978
Tardive dyskinesia	Rat	Kobayashi et al., 1977
Retinal degeneration	W_{ag}/R_{ij} rats	Lai et al., 1980
Hydrocephalus, erythematosus	Dog	Leech et al., 1978
Periodontitis	Rice rat	Leonard, 1979
Osteopetrosis, Albers–Schönberg disease, marble bone disease	Angus and Hereford calves	Leipold and Cook, 1977
Hydrocephalus	Calf	Leech et al., 1978
Infantile spinal muscle atrophy	Wobbler (Wr/Wr) mouse	Leestma, 1980
Chronic destructive periodontitis (periodontal disease, pyorrhea, alveolaris)	Marmosets	Levy, 1976
Systemic lupus erythematosus	Dog	Lewis, 1972
Autoimmune response to acetylcholine receptor in myasthenia gravis	Rat	Lindstrom, 1978
Cholelithiasis	Brazilian squirrel monkey	Lofland, 1975
Carcinoma of pancreas	Wistar and Lewis rats	Longneckar
Aleuka	Cat	Lutsky and Mor, 1981
Cerebral calcification, cytomegalovirus infection	Murine cytomegalovirus-induced	Lussier, 1975
Hyposplenia, asplenia, and immunodeficiency	Hereditary in mice	Machado and Lozzio, 1976
Endometriosis	Rhesus monkeys	MacKenzie and Casey, 1975
Kuru, Creutzfeldt–Jakob disease (slow virus infections)	Transmissible mink encephalopathy, scabie-like disease of mink	Marsh, 1972
Down's syndrome (mongolism, trisomy 21)	Chimpanzee	McClure, 1972
Combined immunodeficiency (severe), Swiss-type αgammaglobulinemia	Horse	McGuire et al., 1975

TABLE 1-2. (Continued)

Disease	*Animal Model*	*Reference*
Hypercalcemia associated with adenocarcinoma of apocrine glands of the anal sac	Dog	Meuten et al., 1982
Malignant lymphoma	Lymphosarcoma in cattle, sheep, horses, and pigs	Miller, 1974
Skeletal osteosarcoma	Dog	Misdorp, 1980
Breast cancer	Cat	Misdorp and Weijer, 1980
Metastasizing mammary tumors	Asplenic mice	Mitchell et al., 1982
Bronchogenic carcinoma	Syrian golden hamster	Mohr and Ketkar, 1980
GABA control of differentiation and development	Mollusk (*Haliotis rufescens*)	Morse et al., 1980
Cardiac physiology	Sea potato tunicate	Morad and Cleemann, 1980
Trypanosomiasis, sleeping sickness	Deer mice	Moulton and Stevens, 1978
Mania	Mice, rats	Murphy, 1977
Ulcerative colitis	Syngeneric adult rats	Nairn et al., 1979
Malignant hyperthermia	Porcine	Nelson et al., 1976
α_1-Antitrypsin deficiency	Round heart disease of turkeys	Neumann et al., 1976
Primary hepatocellular carcinoma	Rat	Newberne and Rogers, 1973
Adenocarcinoma of colon	DMH-induced in rats	Newberne and Rogers, 1973
Bronchiolo-Alveolar cell carcinoma	Pulmonary adenomatosis of sheep	Nobel and Perk, 1978
Renal osteodystrophy	Dog	Norrdin, 1981
Testicular feminization	Mice	Ohno, 1974
Leukemic lymphoma	Cat	Olsen et al., 1980
Hydranencephaly, porencephaly, cerebral cysts, CNS malfunctions	Blue-tongue-vaccine-virus infection in fetal lambs	Osburn and Silverstein, 1972
Wilson's disease	Bedlington terriers	Owen and Ludwig, 1981
Lymphocytic thyroiditis	Rat	Penhale, 1982
Hereditary fructose intolerance	Rat	Phillips et al., 1974
Effect of stimulant and depressant drugs	Great apes, Rhesus monkeys	Pieper, 1976
G_{M2}-gangliosidosis	Porcine cerebrospinal lipodystrophy	Pierce et al., 1976

TABLE 1-2. (Continued)

Disease	*Animal Model*	*Reference*
Metastatic adenocarcinoma of the prostrate	Rat	Pollard, 1977
Alcohol dependence	Mice, rats	Pohorecky, 1981
Acute and chronic lead nephrapathy	Mice	Port, 1976
Bilateral hemispheric ischemia	Rat	Pulsinelli and Brierley, 1979
Lead encephalopathy	Neonatal Long–Evans rats	Press, 1977
Chediak–Higashi syndrome	Mink, cattle, mice, cats, killer whale	Prieur and Collier, 1978
Xenobiotic toxicity and elimination	Teleost	Pritchard and Miller, 1980
Inflammatory bowel disease (ulcerative colitis, Crohn's disease)	New Zealand white rabbits, guinea pig	Rabin, 1980
Benign monoclonal gammopathy	IP and C_{57}BL mice	Radl, 1981
Carcinoids and adenocarcinoma of stomach	Mastomys	Randeria, 1979
Stroke	Surgical ligation of the middle cerebral artery in rat	Robinson, 1981
Oxygen toxicity	Baboon (*Papio* species)	Robinson et al., 1974
Alcohol or nutritional fatty liver and cirrhosis	Lipotrope-deficient mole rats	Rogers and Newberne, 1973
Spontaneous diabetes mellitus	New Zealand white rabbit	Roth and Conaway, 1982
Carcinogenesis of the uterine cervix	Mice	Rubio, 1977
Infantile spinal muscular atrophy, Werdnig–Hoffman disease	Hereditary neuronal abiotrophy in Swedish Lapland dogs	Sandefeldt et al., 1976
Spontaneous atherosclerosis	Pigeon	Santerre et al., 1972
Skeletal hyperostoses		Schmidt and Smith, 1981
Diabetes mellitus	Spontaneous, BB Wistar rat	Seemayer et al., 1980
Multiple exostoses	Horse	Shupe et al., 1981
Pyruvate kinase deficiency	Dog	Searcy et al., 1979
Malignant melanoma	Platyfish	Sobel et al., 1976
Giardiasis	Albino mice	Stevens and Roberts-Thomson, 1978

TABLE 1-2. (Continued)

Disease	*Animal Model*	*Reference*
Alimentary toxic aleukia, fetal brain necrosis, and renal tubular necrosis	Dogs, mice, pigs, rats	Szezech and Hood, 1978
Diabetes mellitus	*Mystromys albicaudatus*	Stuhlman, 1979
Active chronic hepatitis	Caused by reovirus Type 3 in mice	Stanley and Joske, 1975
Chronic biliary obstruction	Caused by reovirus Type 3 in mice	Stanley and Joske, 1975
Rheumatoid arthritis	Arthritis due to mycoplasma in rats and in swine	Sokoloff, 1973
Gross congenital malfunctions	Rat, mouse, golden hamster, guinea pig, rabbit, monkey, pig, dog	Shenefelt, 1972
Inherited erythrocyte glutathone deficiency	Partial γ-glutamylcysene synthetase deficiency in sheep	Smith, 1976
Waterhouse–Friderichzen syndrome	Acrylonitrile-induced adrenal apoplexy	Szabo et al., 1976
Benign cystic and malignant ovarian teratoma	Ovarian tumors in inbred strain Lt/Sv mice	Stevens, 1976
Bacillary dysentery, shigellosis	Rhesus monkey	Takeuchi et al., 1975
Infantile cortical hyperostosis, Caffey–Silverman syndrome	Dog	Thornburg, 1979
Hepatitis	Chimpanzee	Thung et al., 1981
Fetal growth retardation	Uterine vessel ligation in pregnant rat	Wigglesworth, 1972
Gangrene	Turkey	Wight et al., 1981
Diabetes mellitus	Sekote: diabetes of nutritional origin in carp	Wolf, 1976
Myelogenous leukemia	RF mouse	Wolman et al., 1982
Chronic bronchitis	Dog	Wheeldon et al., 1979
Huntington's disease	Rat	Yamamura et al., 1977
Pseudohyperparathyroidism	Hypercalcemia associated with lymphosarcoma in dogs	Yarrington et al., 1977
Hypercalcemia of malignancy	VX-2 Carcinoma of rabbit	Young et al., 1978
Endocardial fibroelastosis and fibroelastosis	Burmese cat	Zook and Paasch, 1982

The need for new animal models continues to receive attention (Cornelius, 1969). The December 1980 Supplement to Volume 101 of the *American Journal of Pathology* was devoted to the proceedings of a workshop on the needs of new animal models of human disease held April 28–29, 1980 (at Bethesda, Maryland). The special issue of the journal is available for $4 from Universities Associated for Research and Education in Pathology, Inc.; 9650 Rockville Pike; Bethesda, MD 20014. Unfortunately, there are many diseases, disorders, and biochemical processes for which an appropriate model is unknown. Specific needs of animal models and the criteria for their development have been noted for the gastrointestinal system (Arias, 1980; Cheville, 1980); the respiratory system (Reid, 1980; Slauson and Hahn, 1980); the endocrine system (Cahill, 1980; Capen, 1980); the eye (Albert, 1980; Aguirre, 1980); the nervous system (Vogel, 1980; Hadlow, 1980); and the reproductive system (Benirschke, 1980; Corbeil, 1980).

Van Citters (1973) provided insight into several animals that could serve as a model for study. One intriguing example is the study of regulation of cerebral circulation in the giraffe. The animal's brain is about 8 ft above its heart, and the hydrostatic equivalent of 8 ft is approximately 200 mm Hg. Van Citters (1973) notes that one can postulate two interesting possibilities that the giraffe may either have essential hypertension or a very bad case of cerebral ischemia when it is standing up. He notes even more exciting possibilities when the animal bends down for a drink or after drinking when he raises his head very suddenly. The peculiar anatomy of the giraffe is compensated for by an equally exaggerated control system of the cerebral circulation.

BRIEF SURVEY OF THE CHAPTERS

This volume contains discussions on the usefulness of various animal models in predicting the distribution of diagnostic agents in humans. Very little information has been collected in reference to the evaluation of radiodiagnostic agents. Pancreas imaging is one area where many potential clinical radiotracers have been developed in small animals but very few have been successful in humans (Risch, 1978). With ^{65}Zn and ^{131}Cs salts concentration of radioactivity was observed in the rat pancreas, but neither could be validated as pancreas imaging agents in larger animals. Labeled amino acids such as 5- and 6-iodotryptophan and iodophenylalanine showed significant levels of radioactivity in the mouse, rat, and dog pancreas (Atkins et al., 1972; Varma et al., 1969; Lambrecht et al., 1974; Counsell et al., 1968), but studies in man were not successful (Taylor and Cottrall, 1973). Using [^{125}I]-*p*-iodophenylalanine, Blomquist (1969) demonstrated that pancreatic specificity is limited to mice, rats, and hamsters but is not found in guinea pigs, rabbits, dogs, cats, squirrel monkeys, frogs, and quail.

In the present volume similar species differences are reported for organ systems such as the heart and the hepatobiliary tract, for blood components such as platelets, and for tumors. Finally, the development of tumor-specific antibodies is discussed in light of species differences. In general the need for a clear understanding of each model is necessary. Animal models are not so much appropriate or not appropriate for a particular study but rather applicable to a specific situation if the difference between the animal model and humans is clearly understood.

A number of attempts have been made to develop radiotracers that follow a specific process in the heart. One such radiotracer is radiolabeled bretylium. This is an especially interesting compound because there is high uptake in the myocardium of the rat, dog, and pig. However, the compound is not taken up in the heart of either male stump tail monkeys or humans. The suggestion is that these two latter species do not possess "uptake two" mechanisms. Recently, ^{99m}Tc cations have been suggested as myocardial imaging agents (Deutsch et al., 1981, 1982). These also showed a distinct species dependence. The cation $^{99m}Tc[dmpe]Cl_2^+$ exhibits high uptake in the heart of the rat, rabbit, dog, and monkey but is not taken up in the myocardium of pigs or humans. In the case of the ^{99m}Tc compounds, no mechanism has been suggested.

In the use of antibodies the concern with animal models is different. There is no question that an antibody to a human antigen will bind to the human antigen. The question addressed in Chapter 3 is not so much a matter of extrapolating from species to species but rather finding a model that validates the antigen and antibody interaction. In the critical review of the tremendous potential and the potential pitfalls, methods that will allow this validation are suggested.

There have been numerous animal tumor models proposed as representative of the human situation. Some have well-defined biochemical characteristics that have been shown to exist in the human. For example, estrogen-dependent breast cancer can be implanted in female rats and the tumors have the characteristic of the human estrogen-dependent breast cancer—a limited number of binding sites that have a high affinity for estrogen. But other aspects such as protein binding, metabolism, and transport may not be the same and therefore the model is never an exact replica of the human situation.

The development of tumor models is made more difficult by the fact that few radiotracers are based on defined biochemical pathways. Most have been designed to take advantage of a physical change in the diseased organ. The chapter on tumor models contains a complete listing of available tumor models and is an ideal reference for investigators interested in developing tumor-specific radiotracers.

Radiolabeling of blood components has been a popular area of investigation in nuclear medicine. Labeled red blood cells have been used for some time for various vascular studies. Labeled leukocytes have been studied as abscess localizing radiotracers. Platelets labeled with ^{111}In have been used for the detection of arterial lesions, coronary artery thrombus, atherosclerosis, and vascular grafts. These uses have been advanced by the group at Washington University. Animal models play a major part in the development of this agent and here again it is especially important to understand the similarities in platelet action between the model and human physiology. The major differences in species occur in platelet response, effects of platelet inhibitors and drugs, and the rate and size of thrombi formation and dissipation.

The hepatobiliary radiopharmaceuticals have been a recent area of interest in ^{99m}Tc radiotracers. One of the difficulties in choosing the radiotracer with the best properties has been the number of different animal models used. The major differences appear when studying the percentage of the injected dose excreted by the hepatobiliary system and pharmacokinetics including liver blood flow and bile flow. This chapter brings together the comparison of hepatobiliary agents and describes the strength and weaknesses of each model.

There has been much concern about using animal data to predict pharmacokinetics in man. The final chapter addresses this problem in general and in specific for renal agents. There are two probable causes for species differences in radiopharmaceutical distribution: (1) differences in metabolic pathways or protein binding in the species compared and (2) differences due to different circulation time and body size. Several solutions are given for the second cause of species differences and these relationships can unify animal and human distribution data. For the particular case of renal physiology no one animal is an ideal representative of human physiology, but the advantages and disadvantages of each are presented.

This book is a first attempt at gathering information on the effect of the choice of animal species on the development of radiopharmaceuticals. We hope it will be the impetus for increasing attention to this important problem.

GUIDELINES FOR RESEARCH WITH ANIMALS

The guiding principles in the care and use of animals as approved by the Council of the American Physiological Society are reproduced with permission from a 1982 issue of the *Journal of the American Physiological Society*.

> Animal experiments are to be undertaken only with the purpose of advancing knowledge. Consideration should be given to the appropriateness of experimental procedures, species of animals used, and number of animals required.
>
> Only animals that are lawfully acquired shall be used in the laboratory, and their retention and use shall be in every case in compliance with federal, state and local laws and regulations, and in accordance with the NIH Guide, 1978.
>
> Animals in the laboratory must receive every consideration for their comfort; they must be properly housed, fed, and their surroundings kept in a sanitary condition.

Appropriate anesthetics must be used to eliminate sensibility to pain during all surgical procedures. Where recovery from anesthesia is necessary during the study, acceptable techniques to minimize pain must be followed. Muscle relaxants or paralytics are not anesthetics and they should not be used alone for surgical restraint. They may be used for surgery in conjunction with drugs known to produce adequate analgesia. Where use of anesthetics would negate the results of the experiment such procedures should be carried out in strict accordance with the NIH Guide. If the study requires the death of the animal, the animal must be killed in a humane manner at the conclusion of the observations.

The postoperative care of animals shall be such as to minimize discomfort and pain, and in any case shall be equivalent to accepted practices in schools of veterinary medicine.

When animals are used by students for their education or the advancement of science, such work shall be under the direct supervision of an experienced teacher or investigator. The rules for the care of such animals must be the same as for animals used for research.

A discussion of the regulatory ramifications of the animal welfare act (Schwindham, 1973), and the NIH animal care guidelines (Conner, 1973; DHEW, revised 1978) are required reading for anyone contemplating research with animals. Fowler (1978) summarizes methodology on the restraint and handling of wild and domestic animals. The volumes on *Methods in Animal Experimentation* (Gay, 1965, 1966, 1968, 1973, 1974) describe fundamental and well-defined techniques for the most efficient use of laboratory research animals and provide guidelines for the welfare and comfort of animals used in various fields of research. Comprehensive bibliographies are provided. Proceedings from symposia on research animals and concepts of applicability to clinical medicine discuss scientifically certain opinions of people involved with problems of animal welfare (Gärtner et al., 1982).

The *Merck Handbook of Diagnosis and Therapy for the Veterinarian* (Siegmund, 1979); the *Handbook of Veterinary Surgical Instruments and Glossary of Surgical Terms* (Hurov, 1978) are recommended desktop references.

The *Directory of Sources of Animals for Research* (National Academy of Sciences, 1979); The *Fascicle of the Handbook of Animal Models* (Registry of Comparative Pathology, 1982); and all articles in the section on Animal Models of Human Disease published in the *American Journal of Pathology* (1972–1982) provide data on the availability of animal models.

ACKNOWLEDGMENT

Research carried out at Brookhaven National Laboratory is under contract with the U.S. Department of Energy and its Office of Health and Environmental Research. Work carried out at George Washington University Medical Center is supported by DHEW Grants HL19127, CA18675, and CA28464 and U.S. Department of Energy Award DOE ER60039.

REFERENCES

Abel, E. L. (1982). *Fetal Alcohol Syndrome*, Vol. 3, *Animal Models*. Boca Raton, Fla.: CRC Press. 200 pp.

Aguirre, G. (1980). Criteria for development of animal models of diseases of the eye. *Am. J. Pathol.* **101:**S187–S200.

Alavi, A., Reivich, M., Greenberg, J., and Wolf, A. P. (1982). Cerebral functional activity mapped with ^{18}F-2-deoxy-2-fluoro-D-glucose. In *Applications of Nuclear and Radiochemistry*, Lambrecht, R. M., and Morcos, N., eds. New York: Pergamon Press, pp. 239–250.

Albert, D. M. (1980). Needs for animal models of human disease of the eye. *Am. J. Pathol.* **101:**S177–S185.

Alexander, N. J., ed. (1979). *Animal Models for Research on Contraception and Fertility*. New York: Harper and Row. 607 pp.

Allen, H. L., and Newton, C. D. (1975). Animal model of human disease: Juvenile rheumatoid arthritis in dog. *Am. J. Pathol.* **81:**699–702.

Andre, F., and Andre, C. (1981). Gastric ulcer disease. *Am. J. Pathol.* **102:**133–135.

Andrews, E. J., Ward, B. C., and Altman, N. H., eds. (1979). *Spontaneous Animal Models of Human Disease*, Vol. II. New York: Academic Press.

Anver, M. R., and Cohen, B. J. (1976). Animal model of human disease: Ulcerative colitis induced in guinea pigs with degraded Larrageanan. *Am. J. Pathol.* **84:**435.

Arias, I. M. (1980). Needs for animal models of human diseases of the gastrointestinal system. *Am. J. Pathol.* **101:**S57–S66.

Ashley, L. M. (1973). Animal model of human disease: Liver cell carcinoma, hepatocellular carcinoma, hepatoma. *Am. J. Pathol.* **72:**345–348.

Atkins, H. C., Christman, D. R., Fowler, J. S., Hauser, W., Hoyte, R. M., Klopper, J. F., Lin, S. S., and Wolf, A. P. (1972). Organic radiopharmaceuticals labeled with isotopes of short half life. V. ^{18}F-Labeled 5- and 6-fluorotryptophan. *J. Nucl. Med.* **13:**713–719.

Baba, N., and von Haam, E. (1972). Animal model for human disease: Adenocarcinoma. *Am. J. Pathol.* **68:**653–656.

Baker, H. J., and Lindsey, J. R. (1974). Animal model of human disease: Feline GM_1 gangliosidosis. *Am. J. Pathol.* **74:**649–652.

Banks, K. L. (1979). Animal model of human disease: Glomerulonephritis, autoimmunity, autoantibody. *Am. J. Pathol.* **94:**443–446.

Bannerman, R. M. (1976). Genetic defects of iron transport. *Fed. Proc.* **35:**2281–2285.

Barnhart, G. R., Jones, M., Ishihara, T., Rose, D. M., Chavez, A. M., and Ferrans, V. J. (1982). Animal model of human disease: Degeneration and calcification of bioprosthetic cardiac valves. *Am. J. Pathol.* **106:**136–139.

Bartosek, I., Guaitani, A., and Pacei, E., eds. (1982). *Animals in Toxicological Research*. New York: Raven Press. 214 pp.

Baskerville, A. (1976). Animal model of human disease: Chronic bronchitis. *Am. J. Pathol.* **82:**237–240.

Bend, J. R., and Weber, L. J. (1980). Aquatic animals as models in biomedical research/comparative pharmacology in aquatic species. *Fed. Proc.* **39:**3183–3187.

Benirschke, K. (1980). Needs for animal models of human disease of the reproductive system. *Am. J. Pathol.* **101:**S229–S240.

Berdanier, C. D. (1976). The BHE strain of rat: An example of the role of inheritance in determining metabolic controls. *Fed. Proc.* **34:**2295–2299.

Berridge, M. J., Rapp, P. E., Treherne, J. E. (1979). Cellular oscillations. *J. Exp. Biol.* **81:**1–306.

Blomquist, L. (1969). Species differences in the accumulation of 4-iodophenylalanine in the exocrine pancreas. *Comp. Biochem. Physiol.* **28:**777–782.

Bolande, R. P. (1975). Animal model of human disease: Hirschsprung's disease, aganglionic or hypoganglionic megacolon. *Am. J. Pathol.* **79:**189–192.

Boorman, G. A., and Hollander, C. F. (1976). Animal model of human disease: Medullary carcinoma of the thyroid. *Am. J. Pathol.* **83:**237–240.

Boorman, G. A., Burek, J. D., and Hollander, C. F. (1977). Animal model of human disease: Carcinoma of the ureter and urinary bladder. *Am. J. Pathol.* **88:** 251–254.

Bornschein, R., Pearson, D., and Reiter, L. (1980). Behavioral effects of moderate lead exposure in children and animal models: Part 2, Animal studies. *CRC Critical Rev. Toxicol.* **8:**101–152.

Bowman, B. H., Barnett, D. R., Carson, S. D., and McNeely, M. C. (1980). Studies of cystic fibrosis utilizing mucociliary activity in oyster gills. *Fed. Proc.* **39:**3195–3200.

Bray, G. A. (1977). Experimental models for the study of obesity. *Fed. Proc.* **36:**137–139.

Bray, G. A. (1977). The Zucker-fatty rat: A review. *Fed. Proc.* **36:**148–153.

Brumback, R. A., Susag, M. E., and Gerst, J. W. (1980). Animal model of human disease: Defective skeletal muscle glucose and/or glycogen metabolism. *Am. J. Pathol.* **101:**241–244.

Brummerstedt, E., Basse, A., Flagstad, T., and Andersen, E. (1977). Animal model of human disease: Acrodermatitis enteropathica, zinc malabsorption. *Am. J. Pathol.* **87:**725–728.

Buchner, E., and Buchner, S. (1981). The deoxyglucose method for insects: Towards electron microscopical resolution. *Eur. Neurol.* **20:**152–156.

Bustad, L. K., Hegreberg, G. A., and Padgett, G. A. (1976). *Naturally Occurring Animal Models for Human Disease:* A bibliography. Washington, D.C.: Institute of Laboratory Animal Resources, National Academy of Sciences.

Buznikov, G. A., Manukhin, B. N., Rakic, L. M., and Turpaev, T. M. (1980). Early sea urchin embryo as a model for the study of prenervous functions of neutrotransmitters. New data. In *Circulatory and Developmental Aspects of Brain Metabolism.* Proceedings of 2nd Int. Symp. on the Pathophysiology and Cerebral Energy Metabolism, September 16–20, 1979. Spatz, M., Mrsulja, B. B., Rakic, L. M., and Lust, W. D., eds. New York: Plenum Press, pp. 361–372.

Cahill, G. F. (1980). Needs for animal models of human diseases of the endocrine system. *Am. J. Pathol.* **101:**S131–S140.

Calabrese, E. J. (1978). Animal model of human disease: Increased sensitivity to polychlorinated biphenyls. *Am. J. Pathol.* **91:**405–407.

Calabrese, E. J. (1978). Animal model of human disease: Increased sensitivity to ozone. *Am. J. Pathol.* **91:**409–411.

Capen, C. C. (1980). Criteria for development of animal models of diseases of the endocrine system. *Am. J. Pathol.* **101:**S141–S145.

Capen, C. C., and Black, H. E. (1974). Animal model of human disease: Medullary thyroid carcinoma, multiple endocrine neoplasia, Sipple's syndrome. *Am. J. Pathol.* **74:**377–380.

Capen, C. C., Hackel, D. B., Jones, T. C., and Migaki, G., eds. (1982). *Tenth Fasicle of the Handbook of Animal Models*. Washington, D.C.: Registry of Comparative Pathology, Armed Forces Institute of Pathology.

Capen, C. C., and Martin, S. L. (1975). Animal model of human disease: Cushing's syndrome, hypercortisolism, Cushing's diseases, Nelson's syndrome. *Am. J. Pathol.* **81**:459–462.

Carlson, K. R. (1977). Narcotic mechanisms and dyskinesias. In *Animal Models in Psychiatry and Neurology*. Hanin, I. and Usdin, E., eds. New York: Pergamon Press, pp. 391–404.

Carter, P. B. (1975). Animal model of human disease: Yersinia enteritis. *Am. J. Pathol.* **81**:703–706.

Chandler, F. W., Jr., and Kraus, S. J. (1976). Animal model of human disease: Gonorrhea. *Am. J. Pathol.* **82**:437–440.

Chandler, F. W., Frenkel, J. K., and Campbell, W. G. (1979). Animal model of human disease: Pneumocystis pneumonia. *Am. J. Pathol.* **95**:571–574.

Cheever, A. W., Kuntz, R. E., Moore, J. A., Bryan, G. T., and Brown, R. R. (1976). Animal model of human disease: Carcinoma of the urinary bladder in schistosoma haematobium infection. *Am. J. Pathol.* **84**:673–676.

Chesney, C. F., and Allen, J. R. (1973). Animal model of human disease: Cardiopulmonary disease. *Am. J. Pathol.* **70**:489–492.

Cheville, N. F. (1980). Criteria for development of animal models of disease of the gastrointestinal system. *Am. J. Pathol.* **101**:S67–S86.

Chryssanthou, C. (1981). Animal model of human disease: Dysbaric osteonecrosis. *Am. J. Pathol.* **103**:334–336.

Cohen, A. S., and Shirahama, T. (1972). Animal model of human disease: Amyloidosis. *Am. J. Pathol.* **68**:441–444.

Cohen, E. A. (1953). Human behavior in the concentration camp. New York: Grosset and Dunlap. 295 pp.

Cohen, S. M., and Friedell, G. H. (1979). Animal model of human disease: Carcinoma of the urinary bladder. *Am. J. Pathol.* **95**:849–852.

Conner, M. (1973). NIH animal care guidelines. In *Research Animals in Medicine* Harminson, L. H., ed. DHEW Publication No. (NIH) 72-333. Washington, D.C.: U.S. Government Printing Office, pp. 1274–1277.

Corbeil, L. B. (1980). Criteria for development of animal models of diseases of the reproductive system. *Am. J. Pathol.* **101**:S241–S254.

Cork, L. C., Griffin, J. W., Adams, R. J., and Price, D. L. (1980). Motor neuron disease: Spinal muscular atrophy and amyotrophic laterial sclerosis. *Am. J. Pathol.* **100**:599–602.

Cornelius, C. E. (1969). Animal models—A neglected resource. *N. Engl. J. Med.* **281**:934–944.

Cornelius, C. E., and Arias, I. M. (1972). Animal model of human disease: Crigler-Najjar syndrome. *Am. J. Pathol.* **69**:369–371.

Cotter, S. M. (1977). Animal model of human disease: Acute lymphoblastic leukemia, aplastic anemia. *Am. J. Pathol.* **87**:265–268.

Counsell, R. E., Smith, T. D., Diguilio, W. and Beierwaltes, W. H. (1968). Tumor Localizing Agents VIII. Radioiodinated Phenylalanine Analogs. *J. Pharm. Sci.* **57**:1958–1961.

Craighead, J. E. (1975). Animal model of human disease: Diabetes mellitus. *Am. J. Pathol.* **78**:537–540.

Crawford, E. J., Friedkin, M., Wolf, A. P., Fowler, J. S., Gallagher, B. M., Lambrecht, R. M., MacGregor, R. R., Shiue, C.-Y., Wodinsky, I., and Goldin, A. (1982). ^{18}F-5-Fluorouridine, A new probe for measuring the proliferation of tissue *in vivo*. *Adv. Enzyme Regulation,* Weber, G., ed. **20:**3–22.

Cummings, J. F., and Haas, D. C. (1972). Animal model of human disease: Idiopathic polyneuritis, Guillain-Barré syndrome. *Am. J. Pathol.* **66:**189–192.

DalCanto, M. C., and Lipton, H. L. (1977). Animal model of human disease: Multiple sclerosis. *Am. J. Pathol.* **88:**497–500.

Damjanov, I. (1980). Animal model of human disease: Yolk sac carcinoma (endodermal sinus tumor). *Am. J. Pathol.* **98:**569–572.

Damjanov, I., and Solter, D. (1976). Animal model of human disease: Teratoma and teratocarcinoma. *Am. J. Pathol.* **83:**241–244.

Dardick, I., Setterfield, G., and Sheinin, R. (1978). Animal model of human disease: Megaloblastic anemia. *Am. J. Pathol.* **93:**849–852.

Debons, A. F., Krimsky, I., Maayan, M. L., Fani, F., and Jimenez, F. A. (1977). Gold thioglucose obesity syndrome. *Fed. Proc.* **36:**143–147.

DeLeon, M. J., Ferris, S. H., George, A. E., Christman, D. R., Fowler, J. S., Gentes, C., Reisberg, B., Gee, B., Emmerich, M., Brodie, J., Kricheff, I. I., Wolf, A. P., and Yonekura, Y. (1983). Positron emission tomography studies of aging and Alzheimer's disease. *Am. J. Neuroradiol.* (in press).

Deutsch, E., Bushong, W., Glavan, K. A., Elder, R. C., Sodd, V. J., Scholz, K. L., Fortman, D. L., Lukes, S. J. (1981). Heart imaging with cationic complexes of technetium. *Science* **214:**85–86.

Deutsch, E., Glavan, K. A., Bushong, W., and Sodd, V. J. (1982). The inorganic chemistry of ^{99m}Tc myocardial imaging agents. In *Applications of Nuclear and Radiochemistry,* Lambrecht, R. M., and Morcos, N. A., eds. New York: Pergamon Press, pp. 139–151.

DHEW Publication No. (NIH) 76-982 (1976). *Animal Models of Thrombosis and Hemorrhagic Diseases.* U.S. Department of Health, Education and Welfare. Workshop on Animal Models of Thrombosis and Hemorrhagic Disease, National Academy of Sciences, 1975.

DHEW Publication No. (NIH) 78-23, Revised (1978). *Guide for the Care and Use of Laboratory Animals.* Office of Science and Health Reports, DRR/NIH, Bethesda.

Dodds, W. J. (1982). The pig model for biomedical research—Report of International Symposium, December 4–6, 1980, Taipei, Taiwan, to be published, *Fed. Proc.* **41:**247–256.

Dodds, W. J., and Wilkins, R. J. (1977). Animal model of human disease: Immune-mediated thrombocytopenia, autoimmune thrombocytopenia, idiopathic thrombocytopenic purpura. *Am. J. Pathol.* **86:**489–491.

Donnelly, W. J. C., and Sheahan, B. J. (1975). Animal model of human disease: G_{M1}-Gangliosidosis Type II. *Am. J. Pathol.* **81:**255–258.

Eckelman, W. C., ed. (1982). In *Receptor Binding Radiotracers,* Vols. 1 and 2. Boca Raton, Fla.: CRC Press.

Edelhauser, H. F., Geroski, D. H., and Stern, M. S. (1980). Glucose metabolism in the cornea and lens in elasmobranchs, teleosts and mammals: Response to thiol-oxidation. *Fed. Proc.* **39:**3213–3221.

Eldefrawi, M. E. (1980). Experimental autoimmune myasthenia gravis: The rabbit as an animal model. *Fed. Proc.* **37:**2823–2827.

Eriksson, K. (1980). *Animal Model of Alcohol Research:* Proceedings, International Conference held in Helsinki, June 4–8, 1979. New York: Academic Press. 496 pp.

Farkas, T., Reivich, M., Alavi, A., Greenberg, J. H., Fowler, J. S., MacGregor, R. R., Christman, D. R., and Wolf, A. P. (1980). Application of [^{18}F]2-deoxy-2-fluoro-D-glucose and positron emission tomography in the study of psychiatric conditions. In *Cerebral Metabolism and Neural Function,* Passonneau, J. V., Hawkins, R. A., Lust, W. D., and Welsh, F. A., eds. Baltimore: Williams and Wilkins, pp. 403–408.

Farkas, T., Ferris, S. H., Wolf, A. P., DeLeon, M. J., Christman, D. R., Reisberg, B., Alavi, A., Fowler, J. S., George, A. E., and Reivich, M. (1982). ^{18}F-2-Deoxy-2-fluoro-D-glucose as a tracer in the positron emission tomographic study of senile dementia. *Am. J. Psychiat.* **139:**352–353.

Farkas, T., Wolf, A. P., Jaeger, J., Brodie, J. D., Christman, D. R., Fowler, J. S., MacGregor, R. R., DeLeon, M. J., deFina, P., Goldman, A., Yonekura, Y., Brill, A. B., Schwartz, M., Logan, J., and Cancro, R. (1983). Regional brain glucose metabolism in the study of chronic schizophrenia. *Arch. Gen. Psychiat.* (in press).

Fenstermacher, J. D. (1980). Comparative aspects of blood-brain exchange. *Fed. Proc.* **39:**3201–3206.

Ferris, S. J., DeLeon, M. J., Wolf, A. P., Farkas, T., Christman, D., Reisberg, B., Fowler, J., MacGregor, R., Goldman, A., George, A. E., and Rampal, S. (1980). Positron emission tomography in the study of aging and senile dementia. *Neurobiol. Aging* **1:**127–131.

Festing, M. F. W., ed. (1979). *Animal Models of Obesity.* Oxford: Oxford University Press. 258 pp.

Fischer, J. E. (1977). Animal models in hepatic coma. In *Animal Models in Psychiatry and Neurology,* Hanin, I., and Usdin, E. ed. New York: Pergamon Press, pp. 385–390.

Fletch, S. M., and Pinkerton, P. H. (1973). Animal model of human disease: Congenital hemolytic anemia. *Am. J. Pathol.* **71:**477–480.

Fletcher, T. F., and Kurtz, H. J. (1972). Animal model of human disease: Globoid cell leukodystrophy, Krabbe's disease. *Am. J. Pathol.* **66:**375–378.

Flyger, V., and Levin, E. Y. (1977). Animal model of human disease: Congenital erythropoietic porphyria. *Am. J. Pathol.* **87:**269–272.

Forsberg, J.-G. (1976). Animal model of human disease: Adenosis and clear-cell carcinomas of vagina and cervix. *Am. J. Pathol.* **84:**669–672.

Fowler, M. E. (1978). *Restraint and Handling of Wild and Domestic Animals.* Ames: Iowa State University Press. 332 pp.

Fowler, J. S., and Wolf, A. P. (1982). *The Synthesis of Carbon-11, Fluorine-18 and Nitrogen-13 Labeled Radiotracers for Biomedical Applications.* Washington, D.C.: National Academy of Sciences, NAS-3201. 124 pp.

Frenkel, J. K. (1977). Animal model of human disease: Adrenal infection, necrosis, and hypercorticism. *Am. J. Pathol.* **86:**749–752.

Gärtner, K., Hackbarth, H., and Stolte, H., eds. (1982). *Research Animals and Concepts of Applicability to Clinical Medicine,* Vol. 7, In *Experimental Biology and Medicine,* 1982, Hannover: Karger. 234 pp.

Gay, W. I., ed. (1965, 1966, 1968, 1973, 1974). *Methods in Animal Experimentation,* Volumes 1–15, 382 pp, 608 pp, 469 pp, 384 pp and 400 pp.

Gelatt, K. N., Gum, G. G., Gwin, R. M., Bromberg, N. M., Merideth, R. E., and Samuelson, D. A. (1981). Animal model of human disease: Primary open angle glaucoma. *Am. J. Pathol.* **102:**292–295.

Gershwin, M. E. (1977). Animal model of human disease: DiGeorge syndrome: Congenital thymic hypoplasia. *Am. J. Pathol.* **89:**809–812.

Gershwin, M. E., and Cooper, E. L. (1978). *Animal Models of Comparative and Developmental Aspects of Immunity and Disease,* Proceedings, International Symposium of American Society of Zoologists, Toronto, December 27–30, 1977. New York: Pergamon Press.

Gertz, E. W. (1973). Animal model of human disease: Myocardial failure, muscular dystrophy. *Am. J. Pathol.* **70:**151–154.

Gill, T. J. (1980). The use of randomly bred and genetically defined animals in biomedical research. *Am. J. Pathol.* **101:**S21–S32.

Gindhart, T. D., and Greenspan, J. S. (1980). Animal model of human disease: Angioimmunoblastic lymphadenopathy with dysproteinemia. *Am. J. Pathol.* **99:**805–808.

Glenn, H. J., ed. (1982). *Biologic Applications of Radiotracers.* Boca Raton, Fla.: CRC Press. 224 pp.

Goddard, J., Lewis, R. M., Armstrong, D. L., and Zeller, R. S. (1980). Moderate, rapidly induced hypertension as a cause of intraventricular hemorrhage in the newborn beagle model. *J. Pediat.* **96:**1057–1060.

Goldberg, A. M., and Silbergeld, E. K. (1977). Animal models of hyperactivity. In *Animal Models in Psychiatry and Neurology,* Hanin I., and Usdin, E., eds. New York: Pergamon Press, pp. 371–384.

Gosselin, S., Capen, C. C., and Martin, S. L. (1978). Animal model of human disease: Hashimoto's thyroiditis. *Am. J. Pathol.* **90:**285–288.

Greenberg, J. H., Reivich, M., Alavi, A., Hand, P., Rosenquist, A., Rintelmann, W., Stein, A., Tusa, R., Dann, R., Christman, D., Fowler, J., MacGregor, R., and Wolf, A. P. (1981). Metabolic mapping of functional activity in human subjects with the ^{18}F-fluorodeoxyglucose technique. *Science* **212:**678–680.

Gregory, R. S., and Jones, J. B. (1977). Animal model of human disease: Amyloidosis. *Am. J. Pathol.* **87:**721–724.

Gropp, A. (1975). Animal model of human disease: Autosomal trisomy, developmental impairment and fetal death. *Am. J. Pathol.* **77:**539–542.

Hadlow, W. J. (1980). Animal model of human disease: Criteria for development of animal models of the nervous system. *Am. J. Pathol.* **101:**S213–S227.

Haines, D. E. (1982). The Leaser Bushbaby (Galago) as an animal model: Selected topics, Boca Raton, Fla.: CRC Press. 336 pp.

Hanin, I., and Usdin, E., eds. (1977). *Animal Models in Psychiatry and Neurology.* New York: Pergamon Press. 499 pp.

Hargis, A. M., and Thomassen, R. W. (1979). Animal model of human disease: Solar keratosis (solar dermatosis, senile keratosis) and solar keratosis with squamous cell carcinoma. *Am. J. Pathol.* **94:**193–196.

Harmison, L. T. (1973). *Research Animals in Medicine.* DHEW Publication No. (NIH) 72-333. Washington, D.C.: U.S. Government Printing Office. 1293 pp.

Haschek, W. M., Brody, A. R., Klein-Szanto, A. J. P., and Witschi, H. (1981). Animal model of human disease: Diffuse interstitial pulmonary fibrosis: Pulmonary fibrosis in mice induced by treatment with butylated hydroxytoluene and oxygen. *Am. J. Pathol.* **105:**333–335.

Haskins, M. E., Jezyk, P. F., Desnick, R. J., and Patterson, D. E. (1981). Animal model in human disease: Mucopolysaccharidosis. VI. Maroteauz-Lamy syndrome, arylsulfatase B deficient mucopolysaccharidosis in siamese cat. *Am. J. Pathol.* **105:**191–193.

Hawkes, J. W. (1980). The effect of xenobiotics on fish tissues: Morphological studies. *Fed. Proc.* **39:**3230–3236.

Hay, J. B. (1982). *Animal Models of Immunological Processes.* New York: Academic Press.

Hayes, K. C. (1974). Animal model of human disease: Hemolytic anemia of premature infants associated with vitamin E deficiency. *Am. J. Pathol.* **77:**123–126.

Hegreberg, G. A. (1975). Animal model of human disease: Ehlers-Danlos syndrome. *Am. J. Pathol.* **79:**383–386.

Hegreberg, G. A., Norton, S. L., and Gorham, J. R. (1975). Animal model of human disease: Muscular dystrophy. *Am. J. Pathol.* **85:**233–236.

Hendricks, J. D., Wales, J. H., Sinnhuber, R. O., Nixon, J. E., Loveland, P. M., and Scanlan, R. A. (1980). Rainbow trout (*Salmo gairdniri*) embryos: A sensitive animal model for experimental carcinogenesis. *Fed. Proc.* **39:**3222–3229.

Henson, J. B., and Gorham, J. R. (1973). Animal model of human disease: Persistent viral infections, immunologically mediated glomerulonephritis and arteritis, dysgammopathies. *Am. J. Pathol.* **71:**345–348.

Heppleston, A. G. (1975). Animal model of human disease: Pulmonary alveolar lipoproteinosis. *Am. J. Pathol.* **78:**171.

Hevesy, G. (1948). *Radioactive Indicators.* New York: Interscience. 556 pp.

Hinton, D. E., Heatfield, B. M., Lipsky, M. M., and Trump, B. F. (1980). Animal model of human disease: Renal tubular carcinoma. *Am. J. Pathol.* **100:** 317–320.

Hinze, H. C. (1974). *Animal Models for the Study of Herpes Virus Associated Malignancy,* Vol. 2. New York: Irvington. 156 pp.

Holt, P. G., Keast, D., and Mackenzie, J. S. (1978). Animal model of human disease: Infectious and neoplastic respiratory diseases associated with cigarette smoking. *Am. J. Pathol.* **90:**281–284.

Holtzman, N. A. (1976). Menkes' kinky hair syndrome: A genetic disease involving copper. *Fed. Proc.* **35:**2276–2280.

Homburger, F., Soto, H., Althoff, J., Dalquen, P., and Heitz, P. (1979). Animal model of human disease: Carcinoma of the larynx in hamsters exposed to cigarette smoke. *Am. J. Pathol.* **95:**845–848.

Hong, C. B., Winston, J. M., and Lee, C. C. (1980). Animal model of human disease: Hepatic angiosarcoma. *Am. J. Pathol.* **101:**737–740.

Hook, R. R., Jr., Berkelhammer, J., and Oxenhandler, R. W. (1982). Animal model of human disease: Melanoma, 1982. *Am. J. Pathol.* **108:**130–133.

Hoyte, R. M., Lin, S. S., Christman, D. R., Atkins, H. L., Hauser, W., and Wolf, A. P. (1971). Organic radiopharmaceuticals labeled with short-lived nuclides. III. ^{18}F-Labeled phenylalanines. *J. Nucl. Med.* **12:**280–286.

Huang, S.-C., Phelps, M. E., Hoffman, E. J., Sideris, K., Selin, C. J., and Kuhl, D. E. (1980). Noninvasive determination of local cerebral metabolic rate of glucose in man. *Am. J. Physiol.* **238:**E69–E82 (Endocrinol. Metab. 1).

Hunt, R. D., and Melendez, L. V. (1974). Animal model of human disease: Burkitt's lymphoma, malignant lymphoma and lymphocytic leukemia. *Am. J. Pathol.* **78:**415–418.

Hurov, L. (1978). *Handbook of Veterinary Surgical Instruments and Glossary of Surgical Terms.* Philadelphia: Saunders. 214 pp.

Hurvitz, A. I. (1980). Animal model of human disease: *Pemphigus vulgaris. Am. J. Pathol.* **98:**861–864.

Ido, T., Wan, C. N., Casella, V., Fowler, J. S., Wolf, A. P., Reivich, M., and Kuhl, D. E. (1978). Labeled 2-deoxy-D-glucose analogs. ^{18}F-Labeled 2-deoxy-2-fluoro-D-glucose, 2-deoxy-2-fluoro-D-mannose and ^{14}C-2-deoxy-2-fluoro-D-glucose. *J. Label. Compounds Radiopharm.* **14:**175–184.

Ingle, D. J., and Shein, H. M. (1980). *Model Systems in Biological Psychiatry.* Boston: MIT Press.

Ito, A. (1976). Animal model of human disease: G_{M2} gangliosidosis. *Am. J. Pathol.* **83:**423–426.

Jervis, H. R., and Takeuchi, A. (1979). Animal model of human disease: Amebic dysentery. *Am. J. Pathol.* **94:**197–200.

John, D.T., and Martinez, A. J. (1975). Animal model of human disease: Central nervous infection with the *Nematodeangiostrongylus cantonensis*. *Am. J. Pathol.* **80:**345–348.

Johnson, A. P., and Taylor-Robinson, D. (1981). Animal model of human disease: Chlamydial genital tract infections: Experimental infection of the primate genital tract with *Chlamydia trachomatis*. *Am. J. Pathol.* **106:**132–135.

Jolly, R. D. (1974). Animal model of human disease: Mannosidosis of children, other inherited lyosomal storage disorders. *Am. J. Pathol.* **74:**211–214.

Jones, T. C. (1980). The value of animal models. *Am. J. Pathol.* **101:**S3–S9.

Jones, T. C., Capen, C. C., Hackel, D. B., and Migaki, G. (1972–1980). Animal model of human disease: Fascicles 1–8, Registry of Comparative Pathology, Armed Forces, Institute of Pathology, Washington, D.C.

Julian, L. M. (1973). Animal model of human disease. Hereditary muscular dystrophy of chickens. *Am. J. Pathol.* **70:**273–276.

Kanarek, R. B., and Hirsch, E. (1977). Dietary-induced overeating in experimental animals. *Fed. Proc.* **36:**154–158.

Kaplow, L. S., and Nadel, E. (1979). Animal model of human disease: Acute lymphoblastic leukemia. *Am. J. Pathol.* **95:**273–276.

Karbe, E. (1973). Animal model of human disease: G_{M2}-gangliosidoses (amaurotic idiocies) Types I, II and III. *Am. J. Pathol.* **71:**151–154.

Karlinsky, J. B., and Snider, G. L. (1978). Animal models of emphysema. *Am. Rev. Respir. Dis.* **117:**1109–1133.

Keenan, K. P., Binn, L. N., and Takeuchi, A. (1979). Animal model of human disease: Acute nonbacterial gastroenteritis. *Am. J. Pathol.* **94:**439–442.

Kentera, D., Susic, D., Machado, E., and Lozzio, B. B. (1977). Animal model of human disease: Hydronephrosis, agenesis of the renal medulla, arterial hypertension. *Am. J. Pathol.* **87:**477–480.

Killam, E. K. (1976). Measurement of anticonvulsant activity in *Papio papio* model of epilepsy. *Fed. Proc.* **35:**2264–2269.

Kitchen, H., Murray, R. E., and Cockrell, B. Y. (1972). Animal model of human disease: Spina bifida, sacral dysgenesis and myelocele. *Am. J. Pathol.* **67:**203–206.

Klawans, H. L., Hitri, A., Nausieda, P. A., and Weiner, W. J. (1977). Animal models of dyskinesia. In *Animal Models in Psychiatry and Neurology*. Hanin, I. and Usdin, E. eds. New York: Pergamon Press. pp. 351–364.

Klausner, J. S., Osborne, C. A., and Griffith, D. P. (1981). Animal model of human disease: Canine struvite urolithiasis. *Am. J. Pathol.* **102:**457–458.

Kobayashi, R. M., Fields, J. Z., Hruska, R. F., Beaumont, K., and Yamamura, H. I. (1977). Brain neurotransmitter receptors and chronic antipsychotic drug treatment: A model for tardive dyskinesia. In *Animal Models in Psychiatry and Neurology*. Hanin, I. and Usdin, E., eds. New York: Pergamon Press, pp. 405–409.

Koestner, A. (1975). Animal model of human disease: Subacute sclerosing panencephalitis, multiple sclerosis. *Am. J. Pathol.* **78:**361–364.

Koletsky, S. (1975). Animal model of human disease: Hypertension, obesity, Type 4 hyperlipidemia, vascular disease. *Am. J. Pathol.* **81:**463–466.

Kramer, J. W. (1981). Animal model of human disease: Inherited early-onset, insulin-requiring diabetes mellitus in Keeshound dogs. *Am. J. Pathol.* **105:**194–196.

Krous, H. F., Altshuler, G., London, W. T., Palmer, A. E., Fucillo, D. A., and Sever, J. L. (1978). Animal model of human disease: Congenital hydrocephalus. *Am. J. Pathol.* **92:**317–320.

Kuhl, D. E., Phelps, M. E., Kowell, A. P., Matter, E. J., Selin, C., and Winter, J. (1980). Effects of stroke on local cerebral metabolism and perfusion: Mapping by emission computed tomography of ^{18}FDG and $^{13}NH_3$. *Ann. Neurol.* **8:**47–60.

Kumar, T. C. A. (1980). *Non-Human Primate Models for Study of Human Reproduction*. Basel: Karger. 232 pp.

Lai, Y.-L., Jacoby, R. O., and Yao, P. C. (1979). Animal model of human disease: Peripheral retinal degeneration. *Am. J. Pathol.* **97:**449–452.

Lai, Y.-L., Jacoby, R. O., Jensen, J. T., and Yao, P. C. (1980). Animal model of human disease: Retinitis pigmentosa. *Am. J. Pathol.* **98:**281–284.

Lambrecht, R. M., Atkins, H. L., Elias, H., Fowler, J. S., Lin, S. S., and Wolf, A. P. (1974). A novel ^{123}I-labeling reagent. XII. Synthesis and loading dose effects of ^{123}I-4-iodophenylalanine and ^{123}I-5- and 6-iodotryptophan. *J. Nucl. Med.* **15:**863–867.

Lambrecht, R. M., and Morcos, N. A., eds. (1982). *Applications of Nuclear and Radiochemistry*. New York: Pergamon Press. 582 pp.

Lambrecht, R. M., and Rescigno, A. (1983). *Tracer Kinetics and Physiologic Modeling-Theory to Practice* in Lecture Notes in Biomathematics. Vol. 48. Springer-Verlag, Heidelberg, 509 pp.

Leader, R. W., and Padgett, G. A. (1980). The genesis and validation of animal models. *Am. J. Pathol.* **101:**S11–S20.

Leech, R. W., Haugse, C. N., and Christoferson, L. A. (1978). Animal model of human disease: Hydrocephalus, congenital hydrocephalus. *Am. J. Pathol.* **92:**567–569.

Leestma, J. E. (1980). Animal model of human disease: Werdnig-Hoffman disease (infantile spinal muscular atrophy). *Am. J. Pathol.* **100:**821–824.

Leipold, H. W., and Cook, J. E. (1977). Animal model of human disease: Osteopetrosis, Albers-Schönberg disease, marble bone disease. *Am. J. Pathol.* **86:**745–748.

Leonard, E. P. (1979). Animal model of human disease: Periodontitis. *Am. J. Pathol.* **96:**643–645.

LePoncin-Lafitte, M. (1982). Ph.D. Thesis, University of Paris. 288 pp.

Levy, B. M. (1976). Animal model of human disease: Chronic destructive periodontitis (periodontal disease, pyorrhea alveolaris, pyorrhea). *Am. J. Pathol.* **83:**637–640.

Lewis, R. M. (1972). Animal model of human disease: Systemic lupus erythematosus. *Am. J. Pathol.* **69:**537–540.

Lindsey, J. R., and Capen, C. C., eds. (1976). Animal models for biomedical research. VI. Metabolic disease. *Fed. Proc.* **35:**1191–1236.

Lindstrom, J. (1978). How the autoimmune response to acetylcholine receptor impairs neuromuscular transmission in myasthemia gravis and its animal model. *Fed. Proc.* **37:**2828–2830.

Lofland, H. B. (1975). Animal model of human disease: Cholelithiasis. *Am. J. Pathol.* **79:**619–622.

Longnecker, D. S. (1981). Animal model of human disease: Carcinoma of the pancreas in azaserine-treated rats. *Am. J. Pathol.* **105:**94–96.

Lussier, G. (1975). Animal model of human disease: Cerebral calcification, cytomegalovirus infection. *Am. J. Pathol.* **80:**555–558.

Lutsky, I. I., and Mor, N. (1981). Animal model of human disease: Alimentory toxic aleukia (septic angina, endemic panmyelotoxicosis, alimentary hemorrhagic aleukia): T-2 toxin-induced intoxication of cats. *Am. J. Pathol.* **104:**189–191.

MacGregor, R. R., Fowler, J. S., Wolf, A. P., Shiue, C.-Y., Lade, R. E., and Wan, C. N. (1981). A synthesis of ^{11}C-2-deoxy-D-glucose for regional metabolic studies. *J. Nucl. Med.* **22:**800–803.

Machado, E. A., and Lozzio, B. B. (1976). Animal model of human disease: Hyposplenia, asplenia, and immunodeficiency. *Am. J. Pathol.* **85:**515–518.

MacKenzie, W. F., and Casey, H. W. (1975). Animal model of human disease: Endometriosis. *Am. J. Pathol.* **80:**341–344.

Maibach, H. I. (1975). *Animal Models in Dermatology: Relevance to Human Dermatopharmacology and Dermatotoxicology.* New York: Churchill Livingstone. 250 pp.

Marsh, R. F. (1972). Animal model of human disease: Kuru, Creutzfeldt-Jakob disease (slow virus infections). *Am. J. Pathol.* **69:**209–212.

Martinez, A. J., Nelson, E. C., and Duma, R. J. (1973). Animal model of human disease: Primary amebic meningoencephalitis, naegleria meningoencephalitis, CNS protozoal infection. *Am. J. Pathol.* **73:**545–548.

Mazziotta, J. C., Phelps, M. E., Carson, R. E., and Kuhl, D. E. (1982). Tomographic mapping of human cerebral metabolism: Auditory stimulation. *Neurology* **32:**921–937.

Mazziotta, J. C., Phelps, M. E., Miller, J., and Kuhl, D. E. (1981). Tomographic mapping of human cerebral metabolism: Normal unstimulated state. *Neurology* **31:**503–516.

McClure, H. M. (1972). Animal model of human disease: Down's syndrome (mongolism, trisomy 21). *Am. J. Pathol.* **67:**413–416.

McGuire, T. C., Banks, K. L., and Poppie, M. J. (1975). Animal model of human disease: Combined immunodeficiency (severe), Swiss-type agammaglobulinemia. *Am. J. Pathol.* **80:**551–554.

Ment, L. R., Stewart, W. B., Duncan, C. C., and Lambrecht, R. M. (1982). Beagle puppy model of intraventricular hemorrhage. *J. Neurosurg.* **57:**219–223.

Meuten, D. J., Capen, C. C., Kociba, G. J., and Cooper, B. J. (1982). Animal model of human disease: Hypercalcemia of malignancy. *Am. J. Pathol.* **108:**366–369.

Miller, J. M. (1974). Animal model of human disease: Malignant lymphoma. *Am. J. Pathol.* **75:**416–420.

Misdorp, W. (1980). Animal model of human disease: Skeletal osteosarcoma. *Am. J. Pathol.* **98:**285–288.

Misdorp, W., and Weijer, K. (1980). Animal model of human disease: Breast cancer. *Am. J. Pathol.* **80:**563–567.

Mitchell, J. R., Lozzio, B. B., Machado, E., Coulson, P., and Linton, E. (1982).

Animal model of human disease. Metastasizing mammary tumors. *Am. J. Pathol.* **108**:249–252.

Mitruka, B. M., Rawnsley, H. M., and Vadherd, D. V. (1976). *Animals for Medical Research: Models for the Study of Human Disease.* New York: Wiley and Sons.

Mohr, U., and Ketkar, M. B. (1980). Animal model of human disease: Bronchogenic carcinoma of the lung. *Am. J. Pathol.* **99**:521–524.

Morad, M., and Cleeman, L. (1980). Tunicate heart as a possible model for the vertebrate heart. *Fed. Proc.* **39**:3188–3194.

Morse, D. E., Duncan, H., Hooker, N., Baloun, A., and Young, G. (1980). GABA induces behavioral and developmental metamorphosis in planktonic mulluscan larvae. *Fed. Proc.* **39**:3237–3241.

Moulton, J. E., and Stevens, D. R. (1978). Animal model of human disease: Trypanosomiasis, sleeping sickness. *Am. J. Pathol.* **91**:693–696.

Murphy, D. L. (1977). Animal models for mania. In *Animal Models in Psychiatry and Neurology.* Hanin, I. and Usdin, E., eds. New York: Pergamon Press, pp. 211–222.

Nairn, R. C., Savvas, R., Hocking, G., Kovala, M., and Rolland, J. M. (1979). Animal model of human disease: Ulcerative colitis. *Am. J. Pathol.* **96**:647–650.

Nathanielsz, P. W. (1980). Animal models in fetal medicine I. In *Monographs in Fetal Medicine,* Vol. 2. New York: Elsevier. 368 pp.

NAS—National Academy of Sciences (1968). *Animal Models for Biomedical Research.* Publication 1736, Washington, D.C.

NAS—National Academy of Sciences (1979). *Animals as Monitors of Environmental Pollutants.* Washington, D.C.

NAS—National Academy of Sciences (1979). *Animals for Research: A Directory of Sources.* Washington, D.C.

NAS—National Academy of Sciences (1981). *Mammalian Models for Research on Aging.* National Academy Press, Washington, D.C. 587 pp.

Navia, J. M. (1977). *Animal Models in Dental Research.* University of Alabama Press, University. 466 pp.

Nelson, T. E., Jones, E. W., and Anderson, I. L. (1976). Animal model of human disease: Malignant hyperthermia. *Am. J. Pathol.* **84**:197–199.

Neumann, F., Meirom, R., Rattner, D., Trainin, Z., Klopfer, U., and Nobel, T. A. (1976). Animal model of human disease: α_1-Antitrypsin deficiency. *Am. J. Pathol.* **84**:427–436.

Newberne, P. M., and Rogers, A. E. (1973). Animal model of human disease: Primary hepatocellular carcinoma. *Am. J. Pathol.* **72**:137–140.

Newberne, P. M., and Rogers, A. E. (1973). Animal model of human disease: Adenocarcinoma of the colon. *Am. J. Pathol.* **72**:541–544.

Nobel, T. A., and Perk, K. (1978). Animal model of human disease: Bronchioloalveolar cell carcinoma. *Am. J. Pathol.* **90**:783–786.

Norrdin, R. W. (1981). Animal model of human disease: Renal osteodystrophy in dogs with radiation. *Am. J. Pathol.* **103**:466–469.

Ohno, S. (1974). Animal model of human disease: Testicular feminization. *Am. J. Pathol.* **76**:589–592.

Okunewick, J. P. (1981). Graft-versus-leukemia in man and animal models. Boca Raton, Fla.: CRC Press. 288 pp.

Olsen, R. G., Mathes, L. E., Hebebrand, L. C., Hoover, E. A., and Nichols, W. S. (1980). Animal model of human disease: Leukemic lymphoma. *Am. J. Pathol.* **98**:857–860.

Osburn, B. I., and Silverstein, A. M. (1972). Animal model of human disease: Hydranencephaly, porencephaly, cerebral cysts, CNS malformations. *Am. J. Pathol.* **67**:211–214.

Owen, C. A., Jr., and Ludwig, J. (1981). Animal model of human disease: Inherited copper toxicosis in Bedlington terriers: Wilson's disease (hepatolenticular degeneration). *Am. J. Pathol.* **106**:432–434.

Panke, T. W., Langlinais, P. C., Vriend, J., and McCue, M. J. (1978). Animal model of human disease: An animal model for childhood convoluted T-cell lymphoma. *Am. J. Pathol.* **92**:595–610.

Penhale, W. J., and Ahmed, S. A. (1982). Animal model of human disease: Lymphocytic thyroiditis. *Am. J. Pathol.* **106**:300–302.

Peterson, H.-I. (1979). Tumor blood circulation: *Angiogenesis, Vascular Morphology and Blood Flow of Experimental and Human Tumors*. Boca Raton, Fla.: CRC Press. 240 pp.

Phelps, M. E. (1981). Positron computed tomography studies of cerebral glucose metabolism in man: Theory and application in nuclear medicine. *Sem. Nucl. Med.* **XI**:32–49.

Phelps, M. E., Huang, S. C., Hoffman, E. J., Selin, C., Sokoloff, L., and Kuhl, D. (1979). Tomographic measurement of local cerebral glucose metabolic rate in humans with (F-18)2-fluoro-2-deoxy-D-glucose; Validation of method. *Ann. Neurol.* **6**:371–388.

Phelps, M. E., Kuhl, D. E., and Mazziotta, J. C. (1981). Metabolic mapping of the brains response to visual stimulation: Studies in humans. *Science* **211**:1445–1447.

Phillips, M. J., Yu, D. T., and Burch, H. B. (1974). Animal model of human disease: Hereditary fructose intolerance. *Am. J. Pathol.* **75**:591–594.

Pieper, W. A. (1976). Great apes and rhesus monkeys as subjects for psychopharmacological studies of stimulants and depressants. *Fed. Proc.* **35**:2254–2257.

Pierce, K. R., Kosanke, S. D., Bay, W. W., and Bridges, C. H. (1976). Animal model of human disease: G_{M2} gangliosidosis. *Am. J. Pathol.* **83**:419–422.

Pohorecky, L. A. (1981). Animal analog of alcohol dependence. *Fed. Proc.* **40**:2056–2064.

Pollard, M. (1977). Animal model of human disease: Metastatic adenocarcinoma of the prostate. *Am. J. Pathol.* **86**:277–280.

Port, C. D. (1976). Animal model of human disease: Acute and chronic lead nephropathy. *Am. J. Pathol.* **85**:519–522.

Press, M. F. (1977). Animal model of human disease: Lead encephalopathy in Children. *Am. J. Pathol.* **86**:484–488.

Prieur, D. J., and Collier, L. L. (1978). Animal model of human disease: Chediak-Higashi syndrome. *Am. J. Pathol.* **91**:533–536.

Pritchard, J. B., and Miller, D. S. (1980). Teleost kidney in evaluation of xenobiotic toxicity and elimination. *Fed. Proc.* **39**:3207–3212.

Prosser, C. L. (1973). Animal models for biomedical research. V. Invertebrates. *Fed. Proc.* **32**:2177–2230.

Pulsinelli, W. A., and Brierley, J. B. (1979). A new model of bilateral hemispheric ischemia in the unanesthetized rat. *Stroke* **10**:267–271.

Purpura, D. P., Penry, J. K., Tower, D. B., Woodbury, D. M., and Walters, R. D., eds. (1972). *Experimental Models of Epilepsy*. New York: Raven Press.

Rabin, B. S. (1980). Animal model of human disease: Immunologic model of inflammatory bowel disease. *Am. J. Pathol.* **99**:253–256.

Radl, J. (1981). Animal model of human disease: Benign monoclonal gammopathy (idiopathic paraproteinemia). *Am. J. Pathol.* **105:**91–93.

Randeria, J. D. (1979). Animal model of human disease: Carcinoids and adenocarcinoma of the stomach. *Am. J. Pathol.* **96:**359–362.

Reid, L. M. (1980). Needs for animal models of human diseases of the respiratory system. *Am. J. Pathol.* **101:**S89–S101.

Reivich, M., Kuhl, D., Wolf, A. P., Greenberg, J., Phelps, M. Ido, T., Fowler, J. S., Hoffman, E., Alavi, A., Som, P., and Sokoloff, L. (1979). The (^{18}F) fluorodeoxyglucose method for the measurement of local cerebral glucose utilization in man. *Circ. Res.* **44:**127–137.

Reivich, M., and Alavi, A. (1983). Pharmacokinetic models and positron emission tomography: studies of physiologic and pathophysiologic conditions. In *Tracer Kinetics and Physiologic Modeling—Theory to Practice,* Lambrecht, R. M., and Rescigno, A., eds. in Lecture Notes in Biomathematics, volume 48, Springer Verlag, Heidelberg 345–383.

Risch, V. R. (1978). Radiopharmaceuticals in pancreatic imaging. In *The Chemistry of Radiopharmaceuticals,* Heindel, N. D., Burns, H. D., Honda, T., and Brady, L. W., eds. New York: Masson Publishing Co.

Robinson, R. G. (1981). Animal model of human disease: A model for the study of stroke using the rat. *Am. J. Pathol.* **104:**103–105.

Robinson, F. R., Casey, H. W., and Weibel, E. R. (1974). Animal model of human disease: Oxygen toxicity. *Am. J. Pathol.* **76:**175–178.

Rogers, A. E., and Newberne, P. M. (1973). Animal model of human disease: Alcoholic or nutritional fatty liver and cirrhosis. *Am. J. Pathol.* **73:**817–820.

Rose, F. C. (1981). *Animal Models of Neurological Disease.* New York: Pitman Books, State Mutual Books. 300 pp.

Roth, S. I., and Conaway, H. H. (1982). Animal model of human disease. Spontaneous diabetes mellitus in the New Zealand white rabbit. *Am. J. Pathol.* **108:**359–363.

Rubio, C. A. (1977). Animal model of human disease: Carcinogenesis in the uterine cervix. *Am. J. Pathol.* **87:**799–802.

Salans, L. B., and Graham, B. J., eds. (1982). Proceedings of a task force on animals appropriate for studying diabetes mellitus and its complications. *Diabetes* **31** (Supplement 1, Part 2):1–102.

Sandefeldt, E., Cummings, J. F., de Lahunta, A., Björck, G., and Krook, L. P. (1976). Animal model of human disease: Infantile spinal muscular atrophy, Werdnig-Hoffman disease. *Am. J. Pathol.* **82:**649–652.

Santerre, R. F., Wight, T. N., Smith, S. C., and Brannigan, D. (1972). Spontaneous atherosclerosis in pigeons: A model system for studying metabolic parameters associated with atherogenesis. *Am. J. Pathol.* **67:**1–22.

Sassenrath, E. N., and Chapman, L. F. (1976). Primate social behavior as a method of analysis of drug action: Studies with THC in monkeys. *Fed. Proc.* **35:**2238–2244.

Schabel, F. M., Jr. (1975). Animal models as predictive systems. In *Cancer Chemotherapy—Fundamental Concepts and Recent Advances.* Chicago: Year Book Medical Publishers, pp. 323–355.

Schmidt, E. V., and Smith, R. E. (1981). Animal model of human disease: Skeletal hyperostoses: Viral induction of avian osteopetrosis. *Am. J. Pathol.* **106:**300–302.

Schmidt-Nielsen, B. (1982). Research animals in experimental medicine. In *Expl.*

Biol. Med., Vol. 7, Gartner, K., Hackbarth, H., and Stolte, H., eds. Hannover: Karger, pp. 46–55.

Schwark, W. S. (1977). Animal model of human disease: Cretinism. *Am. J. Pathol.* **87**:473–476.

Schwartz, E. L., Christman, D. R., and Wolf, A. P. (1983). Human primary visual cortex tomography imaged via positron tomography. *Brain Res.* (in press).

Schwindham, D. F. (1973). Animal Welfare Act (Public Law 91-579) and Amendments in *Research Animals in Medicine,* Harminson, L. H., ed. DHEW Publication No. (NIH) 72-333. Washington, D.C.: U.S. Government Printing Office, pp. 1271–1274.

Shenefelt, R. E. (1972). Animal model of human disease: Gross congenital malformations. *Am. J. Pathol.* **66**:589–592.

Searcy, G. P., Tasker, J. B., and Miller, D. R. (1979). Animal model of human disease: Pyruvate kinase deficiency. *Am. J. Pathol.* **94**:689–692.

Seemayer, T. A., Oligny, L. L., Tannenbaum, G. S., Goldyman, H., and Colle, E. (1980). Animal model of human disease: Diabetes mellitus. *Am. J. Pathol.* **101**:485–488.

Sergan, G., and Kling, A. (1976). *Animal Models in Human Psychobiology.* New York: Plenum Press. 300 pp.

Serio, M., and Martini, L. (1980). *Animal Models in Human Reproduction.* New York: Raven Press. 482 pp.

Shupe, J. L., Leone, N. C., Gardner, E. J., and Olson, A. E. (1981). Animal model of human disease: Hereditary multiple exostoses. *Am. J. Pathol.* **104**:285–288.

Siegmund, D. K., ed. (1979). *The Merck Veterinary Manual: A Handbook of Diagnosis and Therapy for the Veterinarian,* 5th ed. Rahway, N.J.: Merck and Co., Inc.

Slauson, D. O., and Hahn, F. F. (1980). Criteria for development of animal models of diseases of the respiratory system: The comparative approach to respiratory disease model development. *Am. J. Pathol.* **101**:S103–S122.

Smith, J. E. (1976). Animal model of human disease: Inherited erythrocyte glutathione deficiency. *Am. J. Pathol.* **82**:233–235.

Smith, D. W., and Harding, G. E. (1977). Animal model of human disease: Pulmonary tuberculosis. *Am. J. Pathol.* **89**:273–276.

Smith, R. T. (1979). Models for investigation of immune defense mechanism in aquatic animals. *Fed. Proc.* **39**:3184–3187.

Snyder, S. P., and Kingston, R. S. (1982). Animal model of human disease: Niemann-Pick disease. *Am. J. Pathol.* **108**:252–254.

Snyder, R. L., Tyler, G., and Summers, J. (1982). Animal model of human disease: Chronic hepatitis and hepatocellular carcinoma associated with woodchuck hepatitis virus. *Am. J. Pathol.* **107**:422–425.

Sobel, H. J., Marquet, E., Kallman, K. D., and Corley, G. J. (1976). Animal model of human disease: Malignant melanoma. *Am. J. Pathol.* **82**:444.

Sokoloff, L. (1973). Animal model of human disease: Rheumatoid arthritis. *Am. J. Pathol.* **73**:261–264.

Sokoloff, L. (1977). Relation between physiological function and energy metabolism in the central nervous system. *J. Neurochem.* **29**:3–26.

Sokoloff, L., Reivich, M., Kennedy, C., Des Rosiers, M. H., Patlak, C. S., Pettigrew, K. D., Sakurada, O., and Shinohara, M. (1977).The [^{14}C]-deoxyglucose method for the measurement of local cerebral glucose utilization: Theory, procedure, and normal values in the conscious and anesthetized albino rat. *J. Neurochem.* **28:**897–916.

Sokoloff, L., and Smith, C. (1983). Basic principles underlying radioisotopic methods for assay of biochemical processes in vivo. In *Tracer Kinetics and Physiologic Modeling: Theory to Practice,* Lambrecht, R. M., and Rescigno, A., eds. in Lecture Notes in Biomathematics, vol. 48, Springer Verlag, Heidelberg, pp. 202–234.

Spang-Thomsen, M., and Visfeldt, J. (1976). Animal model of human disease: Malignant tumors. *Am. J. Pathol.* **84:**193–196.

Spatz, M., Mrsulja, B. B., Rakic, L. M., and Lust, W. D., eds. (1980). In *Circulatory and Developmental Aspects of Brain Metabolism.* Proceedings of 2nd Int. Symp. on the Pathophysiology and Cerebral Energy Metabolism, September 16–20, 1979. New York: Plenum Press, pp. 361–372.

Spencer, G. R. (1979). Animal model of human disease: Pregnancy and lactational osteoporosis. *Am. J. Pathol.* **95:**277–280.

Stanley, N. F., and Joske, R. A. (1975). Animal model of human disease: Active chronic hepatitis. *Am. J. Pathol.* **80:**181–184.

Stanley, N. F., and Joske, R. A. (1975). Animal model of human disease: Chronic biliary obstruction. *Am. J. Pathol.* **80:**185–187.

Steblay, R. W. (1979). Animal model of human disease: Anti-tubular-basement-membrane-antibody tubulointerstitial nephritis. *Am. J. Pathol.* **96:**649–652.

Steblay, R. W. (1979). Animal model of human disease: Anti-glomerular-basement-membrane glomerulonephritis. *Am. J. Pathol.* **96:**875–878.

Stefanovich, V. (1981). *Animal Models and Hypoxia.* Proceedings of an International Symposium on Animal Models and Hypoxia, held at Wiesbaden, FRG, on November 19, 1979, Advances in Bioscience Series, Vol. 30. New York: Pergamon Press, 138 pp.

Stevens, D. P., and Roberts-Thomson, I. C. (1978). Animal model of human disease: Giardiasis. *Am. J. Pathol.* **91:**529–532.

Stevens, L. C. (1976). Animal model of human disease: Benign cystic and malignant ovarian teratoma. *Am. J. Pathol.* **85:**809–811.

Stinson, S. F. (1979). Animal model of human disease: Esophageal carcinoma. *Am. J. Pathol.* **96:**871–874.

Strandberg, J. D., and Goodman, D. G. (1974). Animal model of human disease: Breast cancer. *Am. J. Pathol.* **75:**225–228.

Storrs, E. E., Binford, C. H., and Migaki, G. (1978). Animal model of human disease: Depromatous leprosy. *Am. J. Pathol.* **92:**813–816.

Stuhlman, R. A. (1979). Animal model of human disease: Diabetes mellitus. *Am. J. Pathol.* **94:**685–688.

Symchych, P. S., and Winchester, P. (1978). Animal model of human disease: Potter's syndrome. *Am. J. Pathol.* **90:**779–782.

Szabo, S. (1978). Animal model of human disease: Duodenal ulcer disease. *Am. J. Pathol.* **93:**273–276.

Szabo, S., Reynolds, E. S., and Kovacs, K. (1976). Animal model of human disease: Waterhouse-Friderichsen syndrome. *Am. J. Pathol.* **82:**653–656.

Szezech, G. M., and Hood, R. D. (1978). Animal model of human disease: Alimentary toxic aleukia, fetal brain necrosis, and renal tubular necrosis. *Am. J. Pathol.* **91:**689–692.

Takeuchi, A., Jervis, H. R., and Formal, S. B. (1975). Animal model of human disease: Bacillary dysentery, shigellosis, shigella dysentery. *Am. J. Pathol.* **81:**251–254.

Tanzer, J. M. (1981). *Animal Models in Cardiology: Symposium Proceedings.* Arlington, Va.: IRL Press. 458 pp.

Taylor, D. M., and Cottrall, M. F. (1973). Evaluation of amine acids labeled with ^{18}F for pancreas scanning. In *Radiopharmaceuticals and Labeled Compounds,* Vol. 1, Vienna: IAEA. pp. 433–441.

Ter-Pogossian, M. M. (1981). Special characteristics and potential for dynamic function studies with PET. Semin. *Nucl. Med.* **11:**12–32.

Ter Pogossian, M. M., Phelps, M. E., Hoffman, E. J., Mullani, N. A., et al. (1975). A positron-emission transaxial tomograph for nuclear imaging (PETT). *Radiology* **114:**89–98.

Thornburg, L. P. (1979). Animal model of human disease: Infantile cortical hyperostosis (Caffey-Silverman syndrome). *Am. J. Pathol.* **95:**575–578.

Thung, S. N., Gerber, M. A., Purcell, R. H., London, W. T., Mihalik, K. B., and Popper, H. (1981). Animal model in human disease: Chimpanzee carriers of hepatitis B virus: Chimpanzee hepatitis B carriers. *Am. J. Pathol.* **108:**328–332.

Valtin, H. (1976). Animal model of human disease: Hereditary hypothalamic diabetes insipidus. *Am. J. Pathol.* **83:**633–636.

Van Citters, R. (1973). The role of animal research in clinical medicine. In *Research Animals in Medicine*. Harmison, L. T., ed. DHEW Publication No. (NIH) 72-333. U.S. Government Printing Office. Washington, D.C. pp. 3–10.

Varma, V. M., Beierwaltes, W. H., Lieberman, L. M., and Counsell, R. E. (1969). Pancreatic concentration of ^{125}I-labeled phenylalanine in mice. *J. Nucl. Med.* **10:**219–223.

Vogel, F. S. (1980). Needs for animal models of human diseases of the nervous system. *Am. J. Pathol.* **101:**S201–S211.

Wheeldon, E. B., Breeze, R. G., and Pirie, H. M. (1979). Animal model of human disease: Chronic bronchitis. *Am. J. Pathol.* **96:**355–358.

Wigglesworth, J. S. (1974). Animal model of human disease: Fetal growth retardation. *Am. J. Pathol.* **77:**347–350.

Wight, P. A. L., Siller, W. G., and Martindale, L. (1981). Animal model of human disease: March gangrene. *Am. J. Pathol.* **103:**159–161.

Wolf, S. (1976). Animal model of human disease: Diabetes mellitus. *Am. J. Pathol.* **85:**805–808.

Wolman, S. R., McMorrow, L. E., and Cohen, M. W. (1982). Animal model of human disease: Myelogenous leukemia in the Rf mouse. *Am. J. Pathol.* **107:**280–284.

Yarrington, J. T., Hoffman, W. and Macy, D. (1977). Animal model of human disease: Pseudohyperparathyroidism. *Am. J. Pathol.* **89:**531–534.

Yamumura, H. I., Hruska, R. E., Schwarcz, R., and Coyle, J. (1977). Effect of striatal kainic acid lesions on muscarinic cholineric receptor binding: Correlation with Huntington's disease. In *Animal Models in Psychiatry and Neurology.* Hanin, I. and Usdin E. eds. New York: Pergamon Press, pp. 411–414.

Young, D. M., Ward, J. M., and Prieur, D. J. (1978). Animal model of human disease: Hypercalcemia of malignancy. *Am. J. Pathol.* **93:**621–622.

Zook, B. C., and Paasch, L. H. (1982). Animal model of human disease: Endocardial fibroelastosis. *Am. J. Pathol.* **106:**435–438.

CHAPTER 2

The Development of Radiotracers that Are Substrates for (Catecholamine) Uptake$_1$ and Uptake$_2$

EDWARD A. CARR, JR.

INTRODUCTION: DEVELOPMENT OF INFORMATIONAL DRUGS

Drugs are often defined as substances administered for diagnostic, prophylactic, or therapeutic purposes. Most of the radioactive substances administered in the practice of nuclear medicine are used for the first purpose and form a large subclass of the class, diagnostic drugs. But the term *diagnostic* is too narrow, for other types of information may also be obtained with this class of drugs and the term *informational drug* seems preferable. Radiopharmaceuticals that permit external detection after administration to man are currently or potentially capable of providing four types of information: the status of a physiologic function (e.g., measurements of iodide uptake by the thyroid gland under various dietary conditions); the presence of an abnormality (e.g., the screening information commonly obtained with scintigrams of various organs); a precise diagnosis [e.g., the nature of a lesion causing hyperaldosteronism (Seabold et al., 1976; Freitas et al., 1979) through combined use of a radioactive and stable drug in a suppression test]; the effect of therapy (e.g., the monitoring of propranolol therapy by serial measurements of cardiac ejection fraction). The first two examples are related to one another but not identical. The third represents a refinement that, it is hoped, will be joined by more examples in the future as we become more adept at the combined use of stable and radioactive drugs. The fourth is a rather crude example of an informational function, monitoring of therapy, that is not yet well developed in nuclear medicine. It is unrealistic to expect each informational drug to provide all four types of information.

In all the preceding the primary objective is to gain information about patients. In contrast to these, studies with radioactive drugs are sometimes conducted in man to learn more about the radioactive substance administered or about a concomitantly administered drug (i.e., they are drug oriented rather than patient oriented). The simplest type of drug-oriented study with a radioactive substance is one in which the primary purpose is to obtain information about absorption, distribution, metabolism, or excretion of the radioactive drug itself, either because the latter is a candidate for subsequent use in nuclear medicine in patient-oriented studies (e.g., as an imaging agent) or because it will subsequently be used in stable form (e.g., as a therapeutic drug) and is first given to man in radioactive form as a convenient way of obtaining kinetic data. In another type of drug-oriented study a radioactive informational drug is given concomitantly with a stable drug in order to obtain either pharmacokinetic or pharmacodynamic information about the latter. This should be distinguished from simple monitoring of therapy. When therapy with a stable drug is monitored by serial use of a radioactive informational drug, there is the underlying assumption that the effect of the stable drug is reasonably well understood. The radioactive drug is then used to learn how well this effect is being accomplished. If, on the other hand, the mechanism underlying the localization of a radioactive drug in a given target is reasonably well understood, then the effect of a concurrently used stable drug on that localization may provide important information about the stable drug. The use of the radioactive iodide ion to study the effect of antithyroid drugs was an early example.

It may be instructive to compare briefly the development of new radioactive informational drugs with the development of new nonradioactive therapeutic drugs. In the first instance candidates are usually chosen for their potential ability to localize in some organ or tissue of interest; they may simply be the radioisotope of some element normally present in high concentration in the area of interest (e.g., iodine), or a radioactive surrogate of such an element (thallium for potassium), or a radiolabeled form of an exogenous compound previously studied in stable form *in vivo* (chlormerodrin), or a new radiolabeled compound chemically related to a stable exogenous compound previously studied *in vivo* (the radioactive compounds related to bretylium and guanethidine discussed later).

In the search for new stable therapeutic drugs a candidate compound is often selected because its chemical structure or some other characteristic provides a reason to believe it will exert a desired effect such as chemotherapy of cancer. But there are some outstanding examples of development of useful drugs by focusing from the start on a well-defined physiologic mechanism (propranolol; cimetidine). The screening of new compounds in the pharmaceutical industry, as currently practiced, is often a blend of these two approaches. But the distinction is not academic. If one knows from the beginning the mechanism whereby a therapeutic drug carries out its desired effect, subsequent clinical use of that drug may be greatly aided. This is especially true if the drug is to be used in combined therapy with other drugs.

Thus the process of developing a new therapeutic drug by starting from a known mechanism is not only intrinsically appealing to investigators but may offer a bonus if the drug reaches the clinic. In an analogous way, a candidate radioactive informational compound is usually selected because some characteristic such as its relation to other compounds with known distribution *in vivo* gives reason to believe the new compound will show a desired localization. But again it is possible to focus from the start on a mechanism and try to develop compounds that will take advantage of this to achieve the desired localization. There may be a bonus if such a compound reaches clinical nuclear medicine, for the informational uses mentioned earlier—especially the assessment of physiologic functions in man—are obviously more reliably performed if one has a clear idea of the mechanisms of localization.

Regardless of the type of drug under development, the ability of studies in animals to predict results in man is greatly affected by the experimental conditions chosen for the animal study. The investigator tries to select doses, schedules, model lesions, etc., that will be relevant to subsequent human experience. But there is one—literally—specific way in which experiments in animals can *never* equal the human situation: The species is *always* different. Therefore the choice of animal species that will predict the effect of a therapeutic drug or the localization of a radioactive drug in man is a major concern of investigators, and the refinement of drug development obtained by focusing on a known mechanism of effect or localization, respectively, does not remove this problem.

To illustrate these general points more precisely the rest of this paper will be concerned with one set of related mechanisms—those involved in the uptake of catecholamines by various tissues. Work from several clinics and laboratories, including our own, will be reviewed in an attempt to answer two questions: Are the processes whereby catecholamines achieve localization in tissues sufficiently well documented to guide radiopharmaceutical research in designing and selecting candidate compounds for eventual use in nuclear medicine, predicting the most likely target tissues and suggesting the most appropriate experimental conditions for preclinical and clinical studies—especially the most appropriate species for preclinical studies? Is such an approach likely to give information of clinical significance if and when all this has been accomplished?

CATECHOLAMINE $UPTAKE_1$ AND $UPTAKE_2$

The sympathetic division of the autonomic (involuntary) nervous system is important in the regulation of many functions of the body, modulating them through impulses to various effector cells, including sites in the heart, various types of smooth muscle, and glands. When the nerve impulse, that is, the action potential, has traveled to the end of a postganglionic (peripheral) sympathetic nerve, transmission of the impulse from the nerve ending to receptors in the effector cell is chemical. The sympathetic nerve endings synthesize norepinephrine (noradrenaline) from precursors and store it in granules. The action potential initiates release of the stored norepinephrine from the nerve ending into the cleft separating it from the effector cell. As postganglionic sympathetic nerve endings, except at a few special sites, release norepinephrine as their transmitter, they are termed adrenergic. Despite this use of a single transmitter substance, the system has selectivity by virtue of differences among receptors. Thus norepinephrine initiates constriction of the smooth muscle of blood vessel walls by binding to α_1 receptors and increases heart rate by binding to β_1 receptors. Two different routes to development of informational drugs, or, indeed, any type of drug that utilizes this system, are thus apparent. A drug may in some way take advantage of the fact that the nerve ending synthesizes, stores, and releases norepinephrine or may take advantage of the selectivity of the adrenergic receptors. But we have still a third string to our bow.

To maintain its fine adjustment of body functions the sympathetic nervous system must be able not only to turn on responses promptly, as in accelerating the heart rate, but also to turn off such responses rapidly when they are no longer appropriate. One way of accomplishing the latter is by metabolizing norepinephrine to less active or inactive substances, and such a mechanism exists. But an even more economical way is re-uptake of the norepinephrine into the sympathetic nerve ending, where the transmitter can again be stored in granules for subsequent use when needed. This mechanism, now termed $uptake_1$ (neuronal uptake), permits both rapid termination of the physiologic response and conservation of the transmitter. This process provides still another possible site for drug action. In view of the importance of localization, that is, selective concentration, in most clinical uses of radioactive drugs, uptake mechanisms are particularly attractive to investigators seeking to develop new radiopharmaceuticals. In addition to synthesis, storage, and release of norepinephrine, the sympathetic nerve ending is capable of intracellular metabolism of norepinephrine and some is destroyed within the nerve ending. Strictly speaking, uptake refers only to influx. The concentration of a compound in a target tissue depends on the balance among influx, transformation in the tissue and efflux, or, if only the concentration of the radionuclide itself is measured, the balance between influx and efflux. Although studies of localization of a radioactive compound provide a useful first approximation of a given uptake process they really show only net uptake.

The medulla of the adrenal gland is packed with granules containing norepinephrine and epinephrine (adrenaline). Under the influence of preganglionic sympathetic nerves, the adrenal medulla can release both these compounds into the blood, permitting them to travel to distant sites and supplement the effect of norepinephrine released from nerve endings. Although $uptake_1$ has a higher affinity for norepinephrine than for epinephrine, the latter and related catecholamines (Fig. 2-1) can also serve as substrates for $uptake_1$.

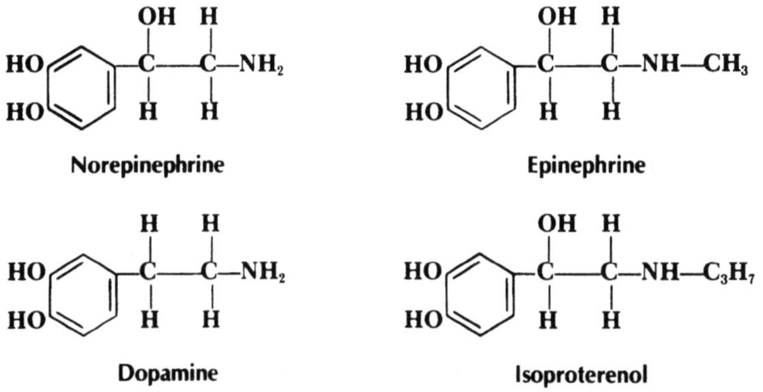

FIG. 2-1. Chemical structures of some important catecholamines.

Although scattered clues existed before 1959, the process now termed uptake$_1$ was first clearly demonstrated by Axelrod et al. (1959), who showed rapid accumulation of [^{3}H]epinephrine in heart, spleen, and other tissues after IV injection into cats. Dengler et al. (1961) showed uptake of [^{3}H]norepinephrine by cat brain slices. Hertting et al. (1961) clearly showed the neuronal nature of this uptake in muscles and glands innervated by postganglionic nerves arising in the superior cervical sympathetic ganglion of cats. These investigators administered [^{3}H]norepinephrine or, in some experiments, [^{3}H]epinephrine to cats that had previously undergone surgical removal of the ganglion on one side. When the muscles and glands of interest were removed and their catecholamine content determined, high uptake of catecholamine was found on the normally innervated side but little uptake was found on the contralateral, denervated side. Subsequently Hamberger et al. (1964) showed uptake of norepinephrine by the adrenergic nerve plexus of the iris of rats. They directly demonstrated concentration of the catecholamine in the nerve tissue by a fluorescence technique. Other tissues with sympathetic innervation that have subsequently been shown to have neuronal uptake include the heart of rat, rabbit, dog and guinea pig, as well as the vas deferens of the rabbit and the cat spleen. Radioautography combined with light microscopy has also been used to demonstrate norepinephrine in some experiments, and sympathetic denervation has been accomplished surgically, by administration of 6-hydroxydopamine or by antibody (immunosympathectomy). Thus the existence of uptake$_1$ in many tissues of many species has been convincingly demonstrated. As far as catecholamine uptake by tissues is concerned, it may seem that all the questions have been answered by these excellent investigations.

In science what happens after all the questions have been answered? At that point all the answers should be questioned. Further work, especially a series of studies by Iversen (Iversen, 1963; Burgen and Iversen, 1965; Iversen, 1965a,b; Iversen and Salt, 1970), has shown that the preceding description, while correct, is incomplete. Using the Langendorff preparation of the isolated heart, Iversen showed that the uptake of norepinephrine and

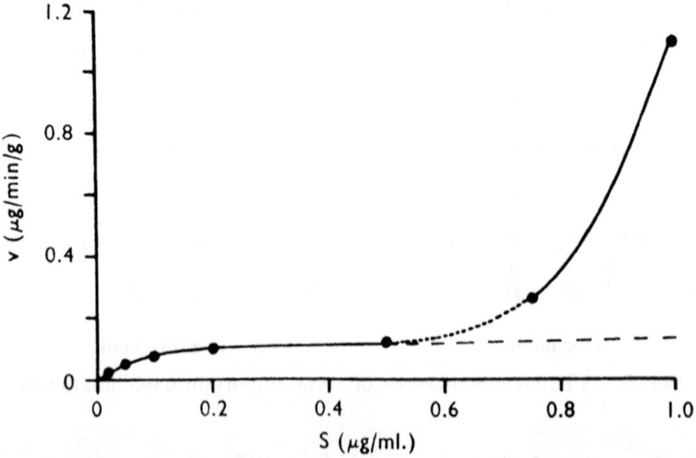

FIG. 2-2. Initial rates of (±)-epinephrine uptake (v, ordinate) by rat isolated heart at various perfusing concentrations (S, abscissa). (Iversen, 1965b). The dashed line shows maximum rate of uptake predicted from a previous study and the values for rates of uptake at concentrations ≤0.5 µg/ml are also from that study (Iversen, 1965a).

epinephrine appeared to be saturated at a concentration of approximately 0.2 µg/ml norepinephrine or 0.5 µg/ml epinephrine, respectively. However, when the concentration of catecholamine in the perfusing fluid was still further increased after the apparent plateau of uptake had been reached, a surprising new and large uptake of catecholamine was observed. Figure 2-2 shows the result with epinephrine. Iversen showed the onset of the second catecholamine uptake process in a dramatic way by measuring uptake of radioactivity in hearts perfused with [^{3}H]epinephrine and then repeating the experiment with addition of various amounts of unlabeled norepinephrine. As the concentration of the latter increased, the uptake of radioactivity decreased, in comparison to controls, in the expected way as the first uptake system was progressively saturated with norepinephrine. But once the concentration of the latter exceeded 0.5 µg/ml, the second uptake system clearly revealed itself (Fig. 2-3). Iversen subsequently concluded that the second uptake system, which he termed $uptake_2$, is operative even at low concentrations of catecholamine, but the low affinity of $uptake_2$ makes the latter difficult to detect at low concentrations, in contrast to the high-affinity $uptake_1$.

$Uptake_2$ has since been found in many types of smooth muscle (vas deferens, spleen, vascular smooth muscle, etc.) and heart. Fluorescence techniques, permitting direct microscopic observation of the localization of catecholamines, have shown that $uptake_2$ is extraneuronal. This demonstration has been accomplished with spleen, arterial smooth muscle, and other tissues. The physiological role of $uptake_2$ remains uncertain. Postulated roles include removal of excess catecholamines, provision of a second pathway to

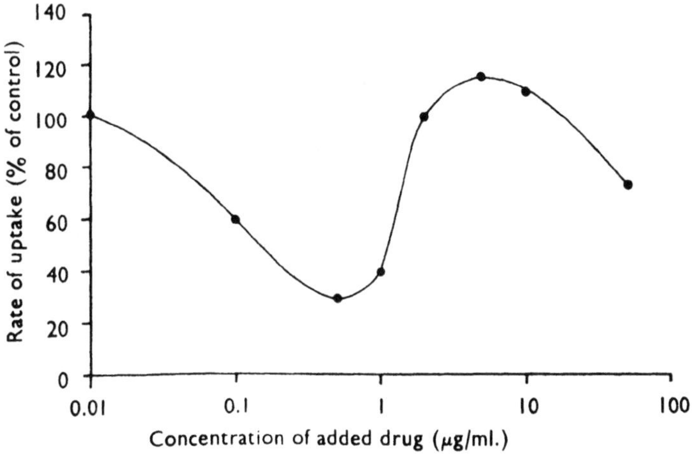

FIG. 2-3. Uptake of [^{3}H]epinephrine in the rat isolated heart perfused with a medium containing a tracer amount of (±)-[^{3}H]epinephrine (2.5 ng/ml) to which various amounts of nonradioactive (±)-norepinephrine were added (Iversen, 1965b).

permit catecholamines to reach tissue sites distant from nerve endings, or initiation of some other as yet unidentified process.

Although the discovery of uptake$_2$ was based on affinity and capacity differences from uptake$_1$, subsequent work has shown several other differences (Table 2-1). An important difference was quickly recognized by Iversen: Several previously known inhibitors of uptake$_1$ are not effective inhibitors of uptake$_2$, which is in turn inhibited by quite a different array of compounds (Figs. 2-4, 2-5). The haloalkylamine, compound SKF 550, is shown in Fig. 2-5 because it is more specific for uptake$_2$ than the more

TABLE 2-1. Characteristics of Uptake$_1$ and Uptake$_2$

Uptake$_1$	*Uptake$_2$*
Neuronal	Extraneuronal
High Affinity	Low Affinity
Low Capacity	High Capacity
Partial Stereospecificity for Norepinephrine	No Stereospecificity for Norepinephrine
Norepinephrine > Epinephrine	Epinephrine > Norepinephrine
Isoproterenol not taken up	Isoproterenol taken up

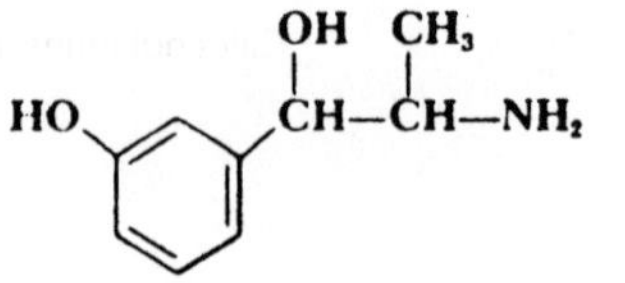

Metaraminol

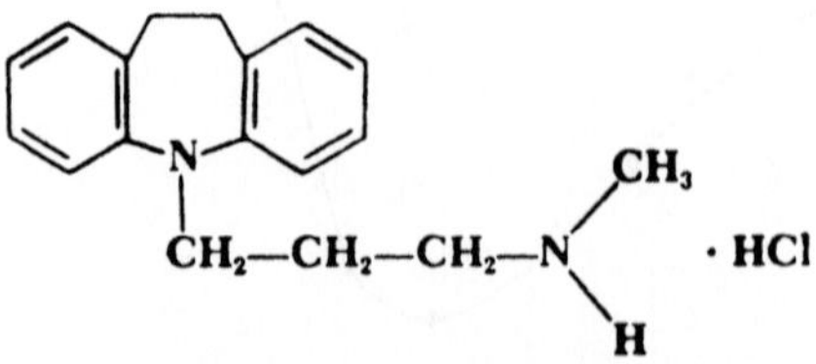

Desipramine hydrochloride

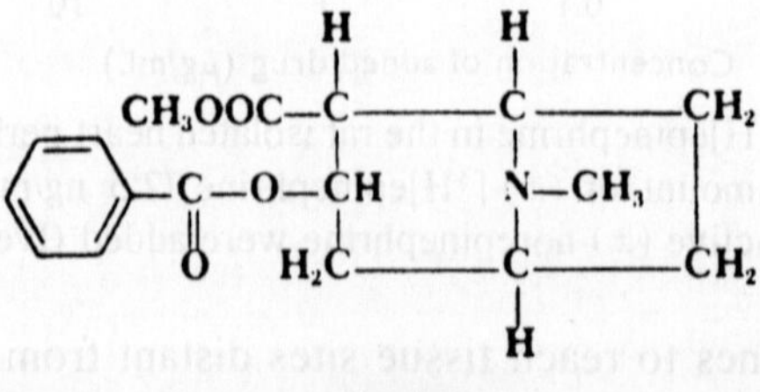

Cocaine

FIG. 2-4. Representative inhibitors of uptake$_1$.

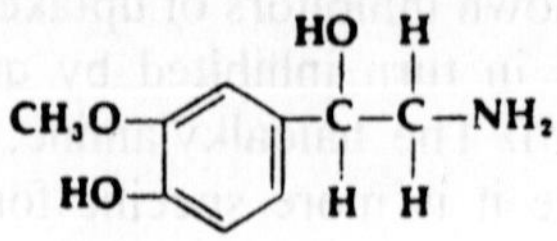

Normetanephrine

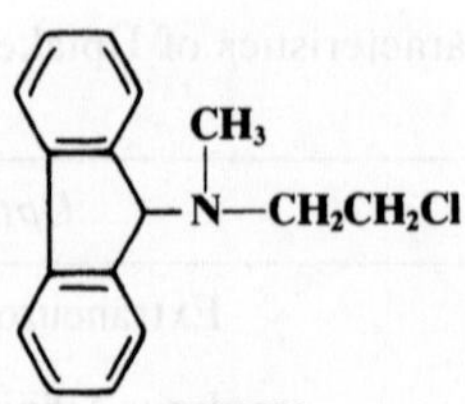

SKF550

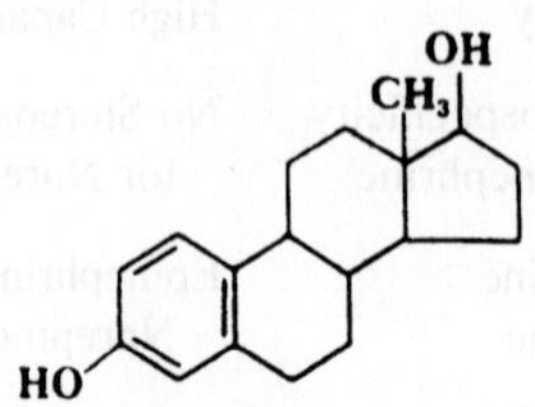

Estradiol

FIG. 2-5. Representative inhibitors of uptake$_2$.

widely known phenoxybenzamine, which inhibits both $uptake_1$ and $uptake_2$ (Iversen et al., 1972). This differing sensitivity to certain inhibitors has been helpful in identifying the mechanism responsible for the uptake of various drugs in animals and especially in human studies.

Although the characterization of $uptake_1$ and $uptake_2$ as neuronal and extraneuronal, respectively, is now firmly established, all selective concentration of catecholamines at sites other than nerve endings is not necessarily via $uptake_2$. This will be discussed later.

The ability of the heart, at least in the rat, to concentrate norepinephrine by both $uptake_1$ and $uptake_2$ suggests that a good myocardial image should be obtainable with $[^{11}C]$norepinephrine. Such imaging has in fact been obtained in the dog heart (Ansari, 1974; Fowler et al., 1974) (Fig. 2-6). Successful imaging does not per se permit a conclusion as to the mechanism in any given species: $uptake_1$, $uptake_2$, or both may be involved. Evidence presented below suggests that the dog heart, like the rat heart, utilizes both mechanisms of catecholamine uptake.

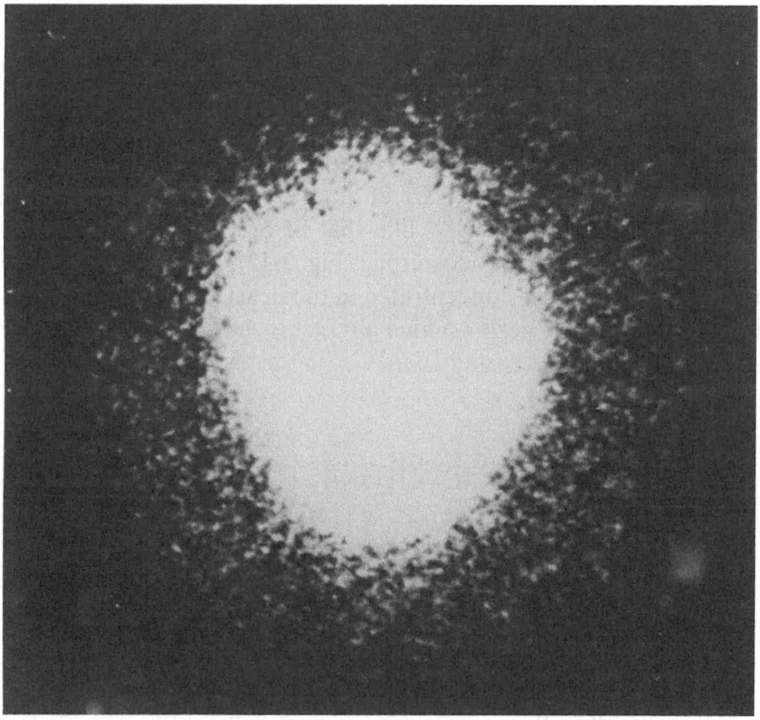

FIG. 2-6. Scintigram of isolated dog heart obtained with a positron camera after injection of $[^{11}C]$norepinephrine (Ansari, 1974; Fowler et al. 1974). (In the recording system used for the scintigram in this and in Figs. 2-13, 2-14, and 2-16, areas of uptake appear white.)

COMPOUNDS OTHER THAN CATECHOLAMINES THAT PARTICIPATE IN UPTAKE$_1$ OR UPTAKE$_2$

Guanethidine and Bretylium (Fig. 2-7)

These compounds were both introduced as antihypertensive drugs. Both interfere with neurotransmission, although their mechanisms of action are not identical. Brodie et al. (1965) found the uptake of guanethidine by rat hearts easily saturated. Cocaine, desipramine, and metaraminol all inhibit it. These findings are consistent with uptake$_1$. Brodie et al. found that the uptake of bretylium by rat heart is not easily saturated. Carr et al. (1979) found that the steroids estradiol and corticosterone, which are potent inhibitors of uptake$_2$ (Iversen and Salt, 1970), inhibit this uptake (Fig. 2-8), and Namm et al. (1975) found that hearts of immunosympathectomized rats have about 80% of normal bretylium uptake. These findings show that the rat heart concentrates bretylium chiefly, not exclusively, by uptake$_2$.

Further Studies with Bretylium

Boura et al. (1960) showed uptake of bretylium by the cat heart. Counsell et al. (1974) subsequently synthesized a series of bretylium analogs. Further study of one analog, *o*-[*I]iodobenzyltrimethylammonium iodide, here given the acronym RIBA (radio-iodinated bretylium analog) (Fig. 2-9), showed good uptake by dog (Figs. 2-10, 2-11) and pig heart but poor uptake by monkey (*M. speciosus*) heart (Carr et al., 1979). In the rat isolated heart preparation the uptake of RIBA, like that of bretylium, was significantly inhibited by estradiol and corticosterone (Fig. 2-12), supporting the view that this analog of bretylium is concentrated in the heart largely by uptake$_2$. The species difference noted earlier added interest to subsequent studies in man.

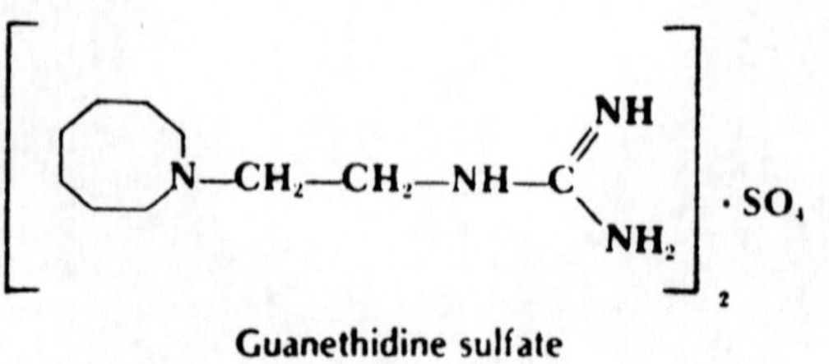

Guanethidine sulfate

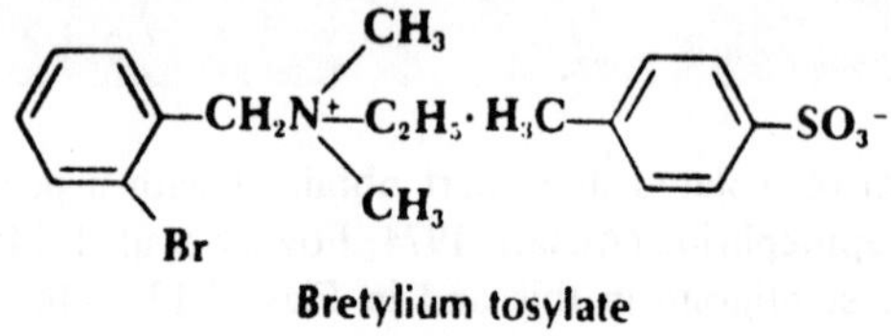

Bretylium tosylate

FIG. 2-7. Chemical structures of guanethidine sulfate and bretylium tosylate.

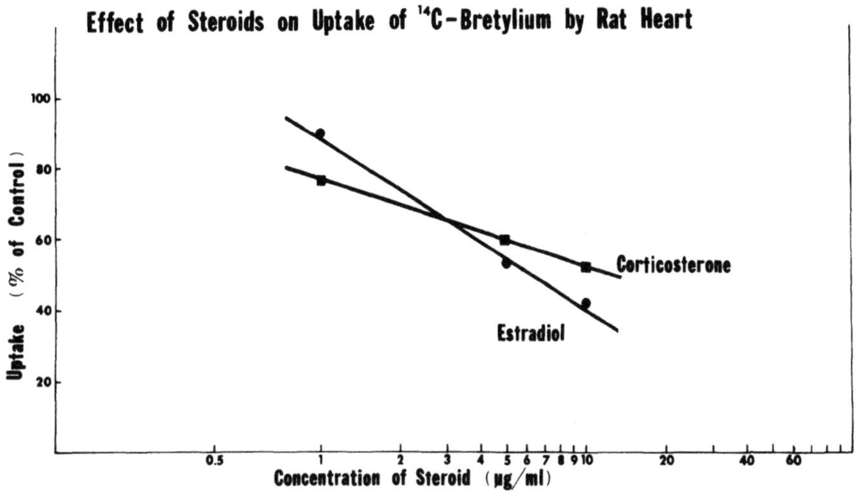

FIG. 2-8. Inhibition of bretylium uptake by steroids. Full data in tabular form are found in Carr et al. (1979).

It was found (Carr et al., 1979) that RIBA, under the conditions of the study, is not taken up significantly by the normal human heart (Figs. 2-13, 2-14). Reasons for believing that this represents a genuine species difference and not a difference in experimental conditions between the animal and human studies were discussed further in the original paper.

Further Studies with Guanethidine

Schanker and Morrison (1965) showed uptake of guanethidine by rat heart after IV administration of 28 mg/kg, but there was also high uptake by skeletal muscle after this dose. As high uptake by intercostal muscles would interfere with the usefulness of a radioactive compound for myocardial imaging, these findings were not encouraging for the development of new

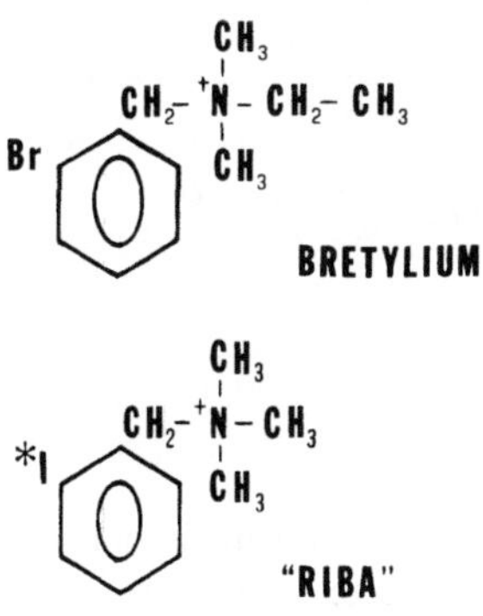

FIG. 2-9. Chemical structures of bretylium and RIBA.

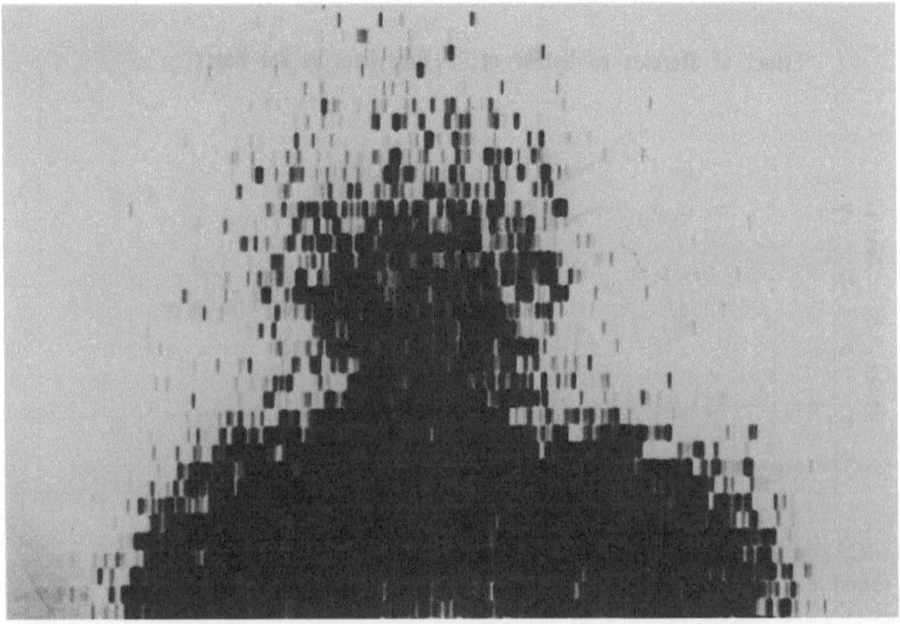

FIG. 2-10. Myocardial scintigram following administration of [^{125}I]RIBA to a normal dog. In this anterior view the heart and liver are seen. (In the recording system used for the scintigram in this and in Fig. 2-11, areas of uptake appear black.)

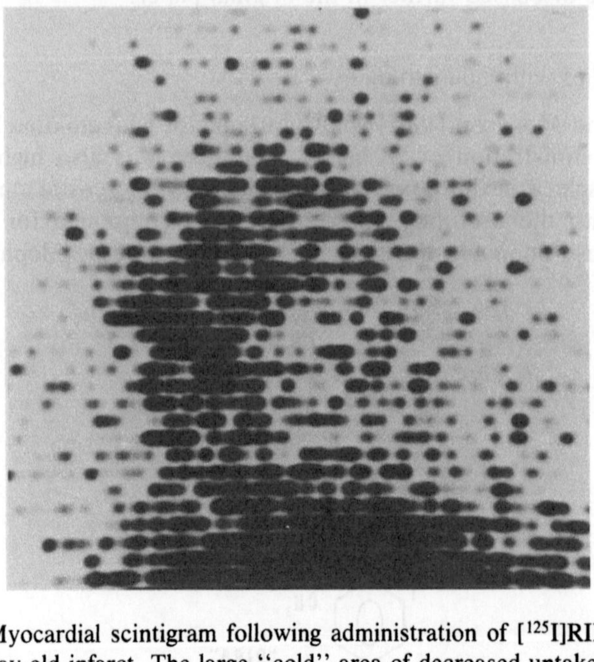

FIG. 2-11. Myocardial scintigram following administration of [^{125}I]RIBA to a dog with a one-day-old infarct. The large "cold" area of decreased uptake seen in this anterior view corresponds to the region of infarction (Counsell et al., 1974).

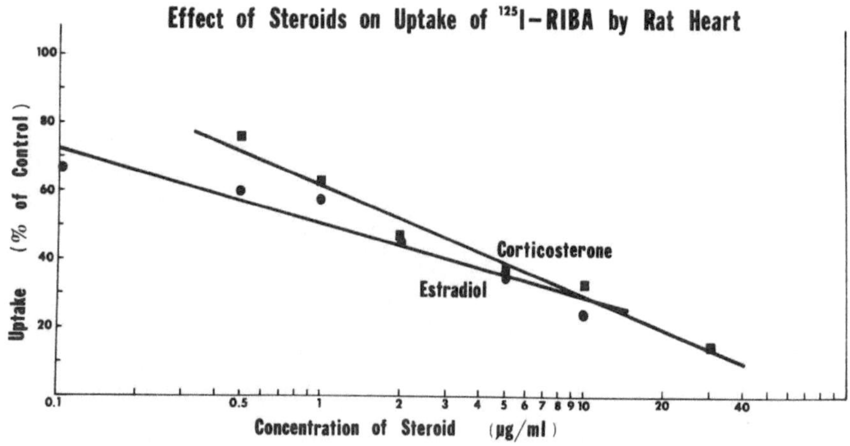

FIG. 2-12. Inhibition of RIBA uptake by steroids. Full data in tabular form are found in Carr et al. (1979).

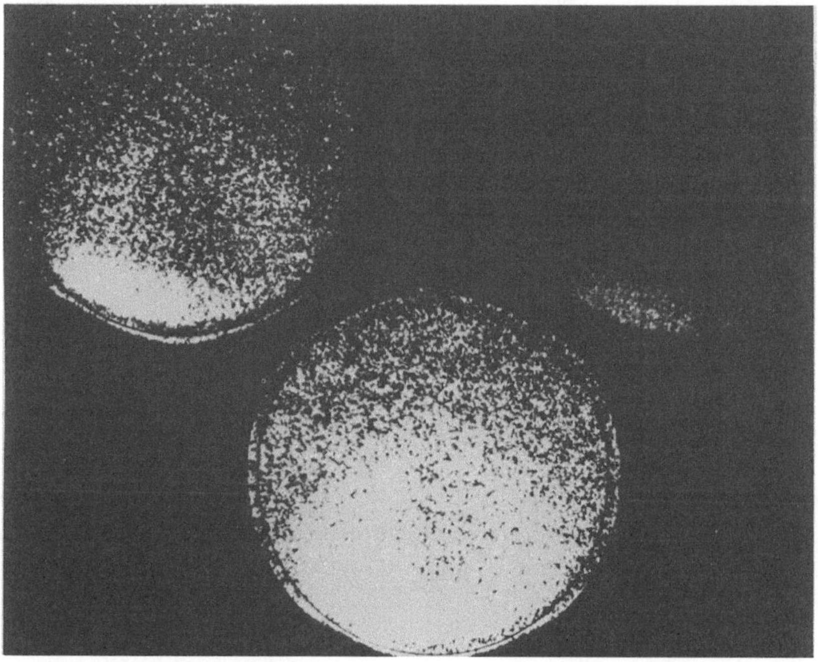

FIG. 2-13. Precordial scintigram obtained in a normal man 110–120 min after IV injection of 10 μCi/kg [^{131}I]RIBA. In this recording system the radiocamera automatically displays a trio of simultaneous images differing in intensity. Some pulmonary density is seen on each side of the heart. The heart itself shows little density. The upper portion of the liver is also seen. (In the recording system used for the scintigram in this and in Figs. 2-14 and 2-16, areas of uptake appear white.)

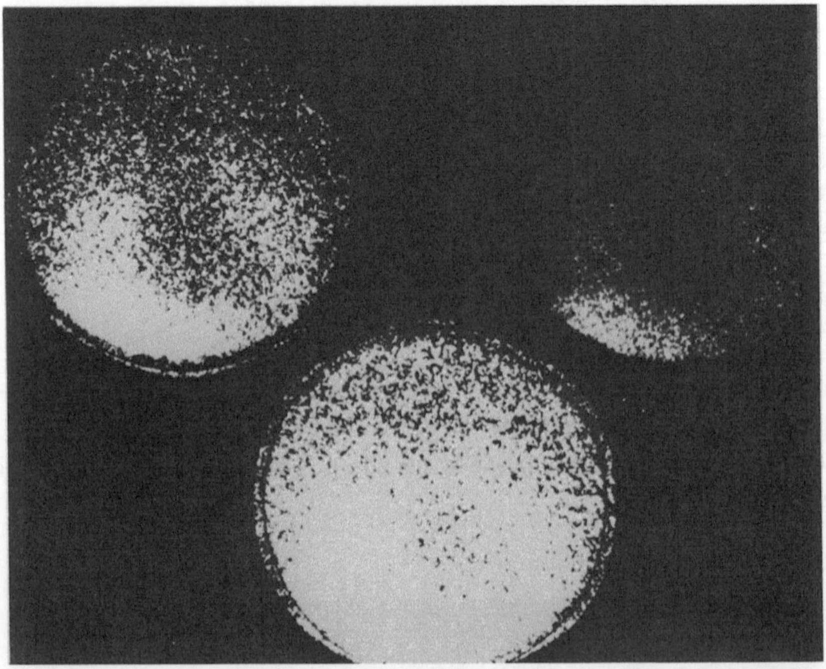

FIG. 2-14. Precordial scintigram obtained in a normal man (not the same subject as in Fig. 2-13) 227–237 min after IV injection of 20 μCi/kg [^{131}I]RIBA. There are definite pulmonary and hepatic densities. There is absence of significant myocardial density (Carr et al., 1979).

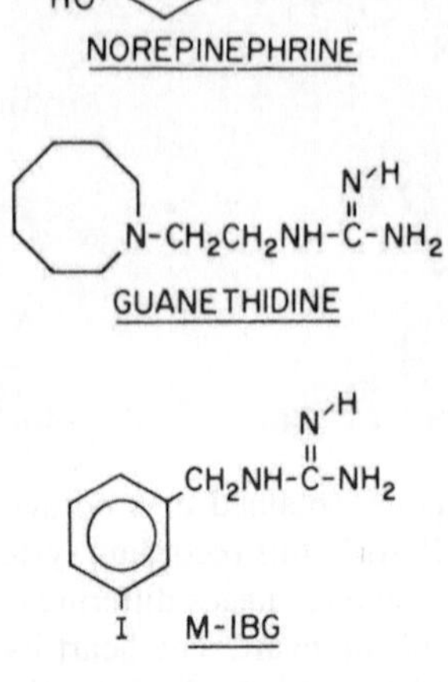

FIG. 2-15. Chemical structures of norepinephrine, guanethidine, and *m*-IBG (Kline et al., 1981).

radioactive compounds structurally related to guanethidine as myocardial imaging agents. But Wieland et al. (1981a) found good uptake of [^{14}C]guanethidine by dog myocardium with little uptake by skeletal muscle at 0.5–6 hours after IV administration of 50 μCi per dog. The specific activity of the administered drug was 0.01 mCi/mg. These investigators also showed high uptake of a related compound, *m*-[^{125}I]iodobenzyl guanethidine (*m*-IBG) (Fig. 2-15) by the heart of rat, dog, and monkey and were able successfully to image the myocardium of the latter two species. Kline et al. (1981) then showed successful imaging of the normal human heart with *m*-IBG (Fig. 2-16).

Guanethidine is a substrate for $uptake_1$. Wieland et al. (1981) showed that myocardial uptake of *m*-IBG at high specific activity is saturable. They also observed slow efflux of radioactivity from the heart after uptake of *m*-IBG. They thus have evidence that *m*-IBG participates in $uptake_1$.

The more exclusively a radioactive informational drug uses one mechanism of $uptake_1$, the more useful will be the information that it provides in the clinic, all other things being equal. It is of interest that Weiland et al.

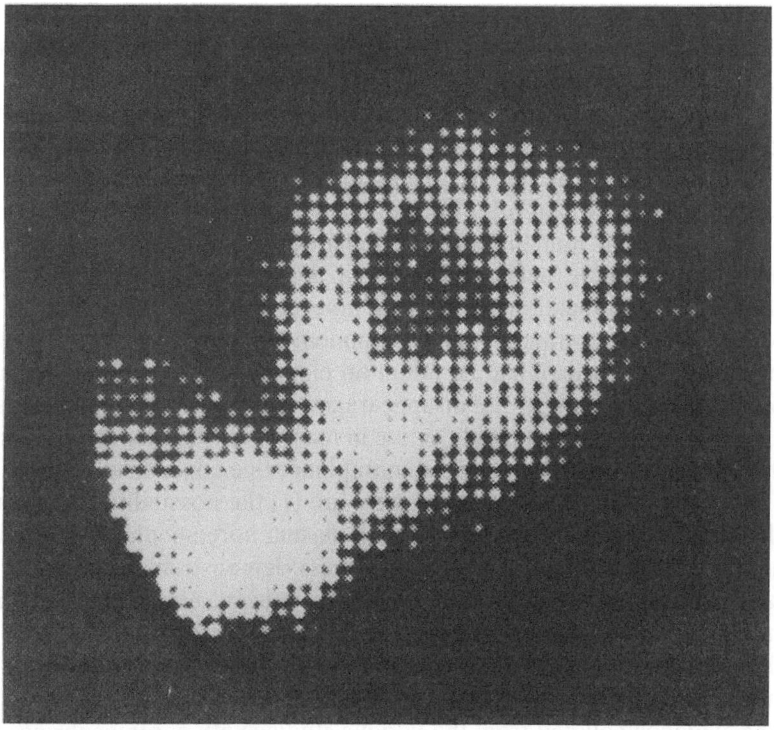

FIG. 2-16. Myocardial scintigram obtained in a normal subject after injection of *m*-[^{125}I]IBG (Kline et al., 1981).

(1981) also found evidence that $uptake_1$ is not the sole mechanism for myocardial uptake of *m*-IBG. Increased uptake of *m*-IBG at low specific activity suggested extraneuronal uptake. As noted earlier, norepinephrine itself is not concentrated in the heart exclusively by $uptake_1$.

SPECIES DIFFERENCES

The studies with RIBA suggested a difference among species with regard to $uptake_2$. Is there any other evidence for differences among species in $uptake_2$? Isoproterenol, a good substrate for $uptake_2$, is taken up very poorly by guinea pig heart (Bönisch and Trendelenburg, 1974). The rabbit also has poor myocardial $uptake_2$ (Bönisch and Trendelenburg, 1974; Lindmar and Löffelholz, 1974), but rabbit aortic strips have good $uptake_2$ (Eckert and Henseling, 1973; Levin, 1974). Thus $uptake_2$, discovered in rat heart, is probably also well developed in cat, dog, and pig heart. Evidence to date suggests that it is poorly developed in the heart of guinea pig, rabbit, at least one species of monkey, and man. The evidence is not equally strong for each species, however, and the list should be considered tentative. Results of fluorescence studies using tissue slices from several species (Gillespie and Muir, 1970) are not fully in accord with results of the perfusion studies just noted, but Gillespie and Muir's results do support the view that different tissues in the same species may differ greatly in $uptake_2$.

The previously mentioned imaging studies are not the only evidence supporting the suggestion that $uptake_2$ may be poorly developed in man. Esler et al. (1981a) showed that the disappearance of [^{3}H]norepinephrine from the plasma of normal humans after infusion to steady state follows a biexponential curve. Normal humans given 125 mg desipramine and patients with peripheral autonomic nerve disease had decreased clearance of norepinephrine and prolongation of the half time of the first component of the curve, strongly suggesting that this component represents $uptake_1$ in man. In contrast, 500 mg cortisol had no effect on plasma norepinephrine clearance or on either component of the disappearance curve in normal humans, suggesting that $uptake_2$ is not highly active in man, at least under the conditions of this experiment where the plasma norepinephrine concentration was 100–300 pg/ml. But three considerations here are (1) the possibility that $uptake_2$ might have been demonstrable at higher plasma norepinephrine concentrations, (2) the counterargument that a process demonstrable only under unusual conditions of very high norepinephrine concentrations may have little physiological significance in any event, and (3) the evidence from animal studies that a species may have poor myocardial $uptake_2$ yet well-developed $uptake_2$ in other tissues.

The evidence collated from the various studies with norepinephrine, bretylium, guanethidine, and related compounds suggests that substrates for $uptake_1$ are better candidates than substrates for $uptake_2$ when the goal is to develop new radioactive informational drugs that will provide their information via myocardial uptake. But all the evidence is by no means in.

THE SPECIAL CASE OF THE ADRENAL MEDULLA

As the adrenal medulla is not nerve tissue, $uptake_1$, strictly defined, does not exist there. But the chromaffin tissue of the adrenal medulla is packed with catecholamine-containing granules similar to those found in adrenergic nerve endings. One might therefore expect, a priori, that any uptake of catecholamines or related compounds by the adrenal medulla would show more similarity to $uptake_1$ than to $uptake_2$. The pharmacologic *effect* of bretylium and guanethidine on the adrenal medulla differs importantly from their effect on adrenergic nerve endings, for both these compounds cause release of norepinephrine from nerve endings, but neither causes release of catecholamines from the granules of the adrenal medulla. Nevertheless, some interesting information on adrenal medullary *uptake* of compounds related to bretylium and guanethidine has been obtained. Korn et al. (1977) found that the para isomer of RIBA concentrates in the adrenal medulla of rats and dogs. Adrenal cortical uptake was much less than medullary uptake, especially several days after administration, when cortical radioactivity had fallen to low levels but high medullary radioactivity persisted (Tables 2-2, 2-3). Although RIBA itself (i.e., the ortho isomer) had shown significant up-

TABLE 2-2. Distribution of ^{125}I in Rat Tissues after Injection of *p*-[^{125}I]RIBA (data abstracted from a more complete table published by Korn et al., 1977)

	[^{125}I] (% Admin [kg] Dose/g) in					
Time p̄ Injection	*Blood*	*Adrenal Medulla*	*Adrenal Cortex*	*Kidney*	*Liver*	*Spleen*
4 days	0.009	0.957	0.166	0.002	0.002	0.028
8 days	<0.001	0.654	0.059	0.001	0.001	0.001

TABLE 2-3. Distribution of ^{125}I in Dog Tissues after Injection of *p*-[^{125}I]RIBA (data abstracted from a more complete table published by Korn et al., 1977)

	[^{125}I] (% Admin [kg] Dose/g) in					
Time p̄ Injection	*Blood*	*Adrenal Medulla*	*Adrenal Cortex*	*Kidney*	*Liver*	*Spleen*
4 days	0.020	0.290	0.008	0.017*	0.018	0.148
8 days	0.012	0.354	0.010	—	0.013	0.134

*48 hrs

take by the whole adrenal gland of rats in a previous study (Counsell et al., 1974), the para isomer had shown greater uptake. Subsequently, Wieland et al. (1979) showed that *o*-[^{131}I]iodobenzyldimethyl-2-hydroxyethylammonium, a compound closely related to RIBA, is taken up well by the adrenal medulla of dogs. This compound permitted imaging of the dog's adrenal medulla four days after injection. It is of interest that neither this compound nor any other bretylium analog has been shown thus far, to our knowledge, to permit satisfactory imaging of the human adrenal medulla. But two compounds related to guanethidine, *m*-IBG and its para isomer, are also taken up by the adrenal medulla (Wieland et al., 1980), and *m*-IBG has been shown to permit imaging of the adrenal medulla of the rhesus monkey (Wieland et al., 1981b) and of the human adrenal medullary tumor, pheochromocytoma, in some instances (Gross et al., 1981; Valk et al., 1981). One is tempted to see in these results another reason for choosing compounds that are substrates for $uptake_1$ over substrates for $uptake_2$ as the starting point in designing new radioactive compounds intended to localize in some human tissue by a catecholamine uptake process. But such a conclusion must be tentative, especially for the adrenal medulla where, as noted earlier, the very terms $uptake_1$ and $uptake_2$ may be misleading.

OTHER EXTRANEURONAL UPTAKE

The myocardium has an active transport system for simple cations such as potassium. Deutsch et al. (1981) synthesized a series of cationic complexes of technetium-99m and showed that the radionuclide, when given in this form, is taken up extensively by the myocardium. It seems more reasonable, in the present state of our knowledge, to relate the mechanism of uptake of Deutsch's complex cations to that of simple cations rather than to catecholamine uptake mechanisms, although the complex cations may conceivably be taken up by some mechanism not related to either. Powis (1973) showed binding of norepinephrine to collagen and elastin and found that various tetracyclines that are not inhibitors of $uptake_1$ or $uptake_2$ inhibit this binding. There is some evidence that another basic compound, tolonium chloride (toluidine blue), and its iodinated derivative are concentrated in the myocardium by an extraneuronal mechanism, although its exact relation to any of the mechanisms discussed earlier is uncertain. Tolonium, after IV administration of 10 mg/kg to dogs, is taken up well by the heart (Kang and DeGiulio, 1968). When [^{125}I]iodotolonium was administered to rats, only modest uptake by heart was obtained (Archer et al., 1969), but when the same radioactive compound was administered to dogs after a priming dose of 10 mg/kg stable tolonium, good myocardial uptake was obtained (Kang, 1969; Carr et al., 1973). Further work clarified the relation of the priming dose of stable tolonium to the absolute myocardial uptake of labeled iodotolonium and to the ratio, [iodotolonium] in myocardium/[iodotolonium] in blood (Ta-

TABLE 2-4. Distribution of ^{131}I in Rats 15 Min after IV Injection of [^{131}I]Iodotolonium (ITB) following a Priming Dose of 10 mg/kg Tolonium (TB) Given as an IV Infusion over Various Time Periods (data abstracted from a more complete table published by Carr et al., 1973)

ITB Uptake in Rats

		Mg/Kg TB Priming Dose		
	Control	*10 in 45′*	*10 in 90′*	*10 in 180′*
L. Ventr.	0.34	1.20*	0.82*	0.78
Liver	0.81	0.62	0.36*	0.32*
Blood	0.14	0.10	0.06*	0.07*

* Signif. ($P < .05$) diff. from control

ble 2-4). This information permitted the use of iodotolonium to obtain successful images of the myocardium in dogs. As iodotolonium is heavily excreted in the bile, the hepatic uptake may simply represent active transport in the excretory process. Assuming the latter to be saturable and capable of transporting both iodotolonium and tolonium, a sufficient priming dose of tolonium should increase the absolute uptake of iodotolonium by many organs, simply by raising the plasma concentration of iodotolonium. But the increase in the myocardial uptake ratio after a priming dose suggests a low-affinity, high-capacity transport process, more characteristic of extraneuronal than of neuronal uptake. The ratio of the radioactivity in the left ventricle to the radioactivity in blood was lower in rabbits than in rats and dogs when all three species received the radioactive compound after a priming dose of 10 mg/kg stable tolonium (Table 2-5). Poor uptake by rabbit heart in comparison to rat and dog heart is suggestive of $uptake_2$, but the absolute uptake of iodotolonium was similar in all three species and the difference in

TABLE 2-5. Distribution of ^{131}I in Dog, Rabbit, and Rat 15 Min after IV Injection of [^{131}I]Iodotolonium (ITB) following a Priming Dose of 10 mg/kg Tolonium Given as an IV Infusion over 45 Min

ITB in 3 Species

		[^{131}I] (% Admin [kg] Dose/g ± S.E.) In		
Species	*N*	*L. Ventr. (V)*	*Blood (B)**	*V/B**
Dog	7	1.48 ± 0.06	0.11 ± 0.01	14.06 ± 1.29
Rabbit	7	1.42 ± 0.21	0.32 ± 0.06	5.38 ± 1.15
Rat	7	1.08 ± 0.25	0.10 ± 0.02	12.20 ± 3.41

* Difference between: rabbit and dog $P < 0.05$; rabbit and rat $P < 0.05$

the ratio was due to the higher radioactivity of blood in rabbits. Therefore the question of uptake$_2$ as a mechanism contributing to iodotolonium uptake by the heart remains open. As these experiments were done primarily to learn whether significant myocardial uptake of iodotolonium could be achieved at all, rather than to investigate its mechanism, inhibition experiments were not performed. Although iodotolonium shows little structural resemblance to catecholamines, the effect of such inhibitors of uptake$_2$ as estradiol and corticosterone would be of interest, as would the effect of tetracyclines. But the use of iodotolonium as an informational drug in man is unfortunately not primising, even if its mechanism of uptake is later established, since electrocardiographic changes have been reported in man after administration of 5 mg/kg IV tolonium over a 2-hour period. Investigation of any low-affinity uptake process poses problems in clinical research because of possible toxic effects from the high doses that may be needed.

POSSIBLE PATHOLOGICAL ROLE OF CHANGES IN UPTAKE$_1$

Is there any evidence that uptake$_1$ is abnormal in any diseased state? Nuzhnyi and Gamelaya (1978) created experimental myocardial hypertrophy in rats by obstructing the aorta and compared uptake of norepinephrine in the isolated hypertrophied hearts with controls, correcting for differences in weight. They found a significant decrease in uptake of norepinephrine by the hypertrophied heart at perfusing concentrations low enough to avoid significant participation of uptake$_2$. They also reported changes in chronotropic and inotropic effects of norepinephrine in the hypertrophied hearts, but these data are more difficult to interpret. Wince and Rutledge (1982) found that uptake$_1$, but not extraneuronal uptake, was significantly decreased in the atria of the offspring of rats fed a diet high in polyunsaturated fats during pregnancy, when compared with atria of offspring of rats fed a diet enriched in saturated fats during pregnancy. The uptake in offspring of dams fed a standard diet fell between the other two groups. In this study each group of rats continued to receive after birth the same diet to which it had been exposed during gestation. In addition to uptake differences, there were differences in chronotropic effect of norepinephrine and isoproterenol, but, as pointed out by the authors, the dietary changes may have influenced the chronotropic response by any of several mechanisms. Moreover, dietary effect on chronotropic response was most marked in adult (65-day-old) rats, while effect on uptake was most marked in younger rats and no longer significant at 65 days.

Reference has already been made to the work of Esler et al. (1981a) in man, in which they produced evidence that the first component of the disappearance curve of norepinephrine from the plasma, after infusion to steady state, represents uptake$_1$. They then showed (Esler et al., 1981b) that this process was impaired in 8 of 32 patients with "essential" hypertension. This abnormality appears to be a stable characteristic of such patients, as it was found in the same patients on subsequent studies. Moreover, the hypertension in some of the patients with apparent abnormality of $uptake_1$ was neither longstanding nor severe, so the effect does not seem to be simply secondary to hypertension itself. One may speculate whether identification of patients with such an abnormality of $uptake_1$, if this can be shown to play a causative or even auxiliary role in hypertension, may have value in predicting response to various antihypertensive drugs, in view of the differences in precise site of action among drugs that have as their common effect interference with adrenergic nerve function.

POSSIBLE FUTURE DIRECTIONS OF RESEARCH

Future directions of research is always a risky topic, for the most significant advances may result from completely unpredicted discoveries. Still it may be useful to consider where current research seems to be carrying us. One can determine if a compound is localized by the mechanism of $uptake_1$; this requires proper choice of reference compound, species, target tissue, inhibitors, and—if further confirmation is still desired—denervation. An important next step may be the use of radioactive informational compounds that are substrates for $uptake_1$ to learn more about the significance of its abnormalities in various human disease states. Although labeled norepinephrine itself is the ideal compound for such studies when this is feasible, as in Esler's work, investigation of $uptake_1$ by individual organs or tissues will require compounds labeled to permit external detection. At present, new candidate $uptake_1$ compounds are probably best studied first in the heart, since myocardial uptake is well characterized, conveniently studied in several laboratory species, and detectable in human scintigrams. Among other target tissues of present interest, the adrenal medulla has the advantage that imaging is feasible and the disadvantage that the status of uptake mechanisms for this tissue is unclear. Arteries have the advantage that their uptake mechanisms are reasonably well understood and the disadvantage that they are not well adapted to present imaging techniques.

Until uncertainty about the status of $uptake_2$ in man is resolved, the most important reason for developing radioactive compounds that localize in appropriate animal tissues by $uptake_2$ is to permit better evaluation of its status in man. Until this question is resolved, questions about possible abnormalities of $uptake_2$ and their clinical significance seem premature. At present, new candidate $uptake_2$ compounds are probably best studied first in the heart, where $uptake_2$ has identifying characteristics analogous to those of $uptake_1$. However, it is already known that $uptake_2$ may be insignificant in the heart of a given species but well developed in some other tissue of that species. The advantages and disadvantages of studying $uptake_2$ in the adrenal medulla and in arteries, respectively, are similar to those for $uptake_1$.

SUMMARY

Radiopharmaceuticals are currently or potentially able to assess functions, screen for abnormalities, establish precise diagnoses, or monitor therapy in man. They thus form a major group of informational drugs. Although there are several routes to the development of new radiopharmaceuticals, the use of a known mechanism of uptake as a starting point offers the possibility not only of a rational progression of preclinical and clinical studies but also the eventual availability of radiopharmaceuticals that will provide reliable information about the status of a given function in man, its abnormalities, and its changes over a period of time. Such information may in some instances guide therapy.

The existence of catecholamine $uptake_1$ (neuronal) is well established in numerous laboratory species and there is good evidence for it in man. As it has a clear physiological role, and there is some evidence that it may be abnormal in disease, radiopharmaceuticals localizing by $uptake_1$ are promising not only as imaging agents per se but as informational drugs in the wider sense. At present new candidate $uptake_1$ radiopharmaceuticals are probably best studied initially in the heart, since cardiac uptake is well characterized, conveniently studied in several laboratory species, and detectable in human scintigrams. Substrate specificity, effect of certain inhibitors, affinity, capacity, and effect of denervation all serve to characterize $uptake_1$. Guanethidine and some of its derivatives appear to localize in the heart chiefly by $uptake_1$.

Catecholamine $uptake_2$ (extraneuronal) is well developed in certain laboratory animals, but poorly developed—at least in the heart—in others, such as rabbit and guinea pig. Its status in man is still uncertain. An important objective in developing radioactive compounds localizing by $uptake_2$ in appropriate laboratory species is to permit subsequent better evaluation of the status of $uptake_2$ in man. At present, new compounds are probably best studied initially in the heart. Identifying characteristics analogous to those for $uptake_1$ exist. Bretylium and some of its derivatives appear to localize in the heart chiefly by $uptake_2$.

Clearly, all extraneuronal uptake of radiopharmaceuticals is not $uptake_2$. Even catecholamines appear to have at least one additional mechanism of extraneuronal uptake. A radiopharmaceutical that localizes in its target organ or tissue exclusively by one well-defined mechanism will yield information that is easier to interpret than information from a radiopharmaceutical that uses more than one mechanism. Many presently known compounds are substrates for more than one catecholamine uptake mechanism, even in a single tissue. Also, a given compound may use $uptake_1$ for localization in one tissue and $uptake_2$ in another. A given species may lack significant $uptake_2$ in one organ but utilize it in others. The low affinity of $uptake_2$ requires special attention to the toxicity of candidate compounds. All these considerations modify, but should not destroy, the hopes expressed at the beginning of this chapter.

REFERENCES

Ansari, A. N. (1974). Myocardial imaging with [^{11}C]-norepinephrine. In *Cardiovascular Nuclear Medicine,* eds. Strauss, H. W., Pitt, B., and James, A. E., Jr., eds. St. Louis: C. V. Mosby, pp. 234–240.

Archer, E. G., Potchen, E. J., Studer, R., and Siegel, B. (1969). Distribution of ^{125}I-toluidine blue in the rat. *J. Nucl. Med.* **10:**386.

Axelrod, J., Weil-Malherbe, H., and Tomchick, R. (1959). The physiological disposition of H^3-epinephrine and its metabolite, metanephrine. *J. Pharmacol. Exp. Ther.* **127:**251–256.

Bönish, H., and Trendelenburg, U. (1974). Extraneuronal removal, accumulation and *O*-methylation of isoprenaline in the perfused heart. *Naunyn Schmiedebergs Arch. Pharmacol.* **283:**191–218.

Boura, A. L. A., Copp, F. C., Duncombe, W. G., Green, A. F., and McCoubrey, A. (1960). The selective accumulation of bretylium in sympathetic ganglia and their postganglionic nerves. *Brit. J. Pharmacol.* **15:**265–270.

Brodie, B. B., Chang, C. C., and Costa, E. (1965). On the mechanism of action of guanethidine and bretylium. *Brit. J. Pharmacol.* **25:**171–178.

Burgen, A. S. C., and Iversen, L. L. (1965). The inhibition of noradrenaline uptake by sympathomimetic amines in the rat isolated heart. *Brit. J. Pharmacol.* **25:**34–39.

Carr, E. A., Jr., Carroll, M., DiGiulio, W., and Blair, D. C. (1973). The use of radioiodinated toluidine blue for myocardial scintigrams. *Am. Heart J.* **86:**631–643.

Carr, E. A., Jr., Carroll, M., Counsell, R. E., and Tyson, J. W. (1979). Studies of uptake of the bretylium analogue, iodobenzyltrimethylammonium iodide, by non-primate, monkey and human hearts. *Brit. J. Clin. Pharmacol.* **8:**425–432.

Counsell, R. E., Yu, T., Ranade, V. V., Buswink, A. A., Carr, E. A., Jr., and Carroll, M. (1974). Radioiodinated bretylium analogs for myocardial scanning. *J. Nucl. Med.* **15:**991–996.

Dengler, H. J., Spiegel, H. E., and Titus, E. O. (1961). Uptake of tritium labelled norepinephrine in brain and other tissues of cat *in vitro*. *Science* **133:**1072–1073.

Deutsch, E., Glavan, K., Sodd, V. J., Nishiyama, H., Ferguson, D. L., and Lukes, S. J. (1981). Cationic Tc-99m complexes as potential myocardial imaging agents. *J. Nucl. Med.* **22:**897–907.

Eckert, E., and Henseling, M. (1973). The effect of corticosterone on extraneuronal uptake in, and efflux of ^{3}H-(±)-noradrenaline from rabbit aortic strips. *Naunyn Schmiedebergs Arch. Pharmacol.* **277:**R14.

Esler, M., Jackman, G., Leonard, P., Skews, H., Bobik, A., and Korner, P. (1981a). Effect of norepinephrine uptake blockers on norepinephrine kinetics. *Clin. Pharmacol. Ther.* **29:**12–20.

Esler, M., Jackman, G., Bobik, A., Leonard, P., Kelleher, D., Skews, H., Jennings, G., and Korner, P. (1981b). Norepinephrine kinetics in essential hypertension. Defective neuronal uptake of norepinephrine in some patients. *Hypertension* **3:**149–156.

Fowler, J. S., Wolf, A. P., Christman, D. R., MacGregor, R. R., Ansari, A., and Atkins, H. L. (1974). Carrier-Free ^{11}C-Labeled Catecholamines. In *Radiopharmaceuticals,* Subramanian, G., Rhodes, B. A., Cooper, J. F., and Sodd, V. J., eds. *Soc. Nucl. Med. New York,* pp. 196–204.

Freitas, J. E., Grekin, R. J., Thrall, J. H., Gross, M. D., Swanson, D. P., and

Beierwaltes, W. H. (1979). Adrenal imaging with iodomethylnorcholesterol (I-131) in primary aldosteronism. *J. Nucl. Med.* **20:**7–10.
Gillespie, J. S., and Muir, T. C. (1970). Species and tissue variation in extraneuronal and neuronal accumulation of noradrenaline. *J. Physiol.* (*Lond.*) **206:**591–604.
Gross, M., Frager, M., Valk, T., Kline, R., Sisson, J., Swanson, D., Wieland, D., Thompsom, N., Topes, M., Beierwaltes, W. (1981). Localization of pheochromocytomas with I-131-M-iodobenzylguanidine (I-131-MIBG). *J. Nucl. Med.* **22:**P5.
Hamberger, B., Malmfors, T., Norberg, K.-A., and Sachs, C. (1964). Uptake and accumulation of catecholamines in peripheral adrenergic neurons of reserpinized animals studied with a histochemical method. *Biochem. Pharmacol.* **13:**841–844.
Hertting, G., Axelrod, J., Kopin, I. J., and Whitby, L. G. (1961). Lack of uptake of catecholamines after chronic denervation of sympathetic neurons. *Nature* **189:**66.
Iversen, L. L. (1963). The uptake of noradrenaline by the isolated perfused rat heart. *Brit. J. Pharmacol.* **21:**523–537.
Iversen, L. L. (1965a). The uptake of adrenaline by the isolated rat heart. *Brit. J. Pharmacol.* **24:**387–394.
Iversen, L. L. (1965b). The uptake of catecholamines at high perfusion concentrations in the rat isolated heart: a novel catecholamine uptake process. *Brit. J. Pharmacol.* **25:**18–33.
Iversen, L. L., and Salt, P. J. (1970). Inhibition of catecholamine uptake$_2$ by steroids in the isolated rat heart. *Brit. J. Pharmacol.* **40:**528–530.
Iversen, L. L., Salt, P. J., and Wilson, H. A. (1972). Inhibition of catecholamine uptake in the isolated rat heart by haloalkylamines related to phenoxybenzamine. *Brit. J. Pharmacol.* **46:**647–657.
Kang, G. S. (1969). Potential value of toluidine blue analogs as scinti scanning agents for myocardial infarcts. *J. Nucl. Med.* **10:**413.
Kang, G. S., and DiGiulio, W. (1968). Potential value of toluidine blue analogs as parathyroid scanning agents. *J. Nucl. Med.* **9:**643–644.
Kline, R. C., Swanson, D. P., Wieland, D. M., Thrall, J. H., Gross, M. D., Pitt, B., and Beierwaltes, W. H. (1981). Myocardial imaging in man with I-123 meta-iodobenzylguanidine. *J. Nucl. Med.* **22:**129–132.
Korn, N., Buswink, A., Yu, T., Carr, E. A., Jr., Carroll, M., and Counsell, R. E. (1977). A radioiodinated bretylium analog as a potential agent for scanning the adrenal medulla. *J. Nucl. Med.* **18:**87–89.
Levin, J. A. (1974). The uptake and metabolism of ^{3}H-1- and ^{3}H-d1-norepinephrine by intact rabbit aorta and by isolated adventitia and media. *J. Pharmacol. Exp. Ther.* **190:**210–226.
Lindmar, R., and Löffelholz, K. (1974). Neuronal and extraneuronal uptake and efflux of catecholamines in the isolated rabbit heart. *Naunyn Schmiedebergs Arch. Pharmacol.* **284:**63–92.
Namm, D. H., Wang, C. M., El-Sayad, S., Copp, F. C., and Maxwell, R. A. (1975). Effects of bretylium on rat cardiac muscle: The electrophysiological effects and its uptake and binding in normal and immunosympathectomized rat hearts. *J. Pharmacol. Exp. Ther.* **193:**194–208.
Nuzhnyi, V. P., and Gamaleya, A. A. (1978). Neuronal and extraneuronal uptake of noradrenaline and the condition of adrenoreactive structures of the rat isolated heart with compensatory myocardial hypertrophy. *Kardiologiia.* **18:**86–91.
Powis, G. (1973). Binding of catecholamines to connective tissue and the effect upon

the responses of blood vessels to noradrenaline and to nerve stimulation. *J. Physiol.* (*Lond.*) **234:**145–162.

Schanker, L. S., and Morrison, A. S. (1965). Physiological disposition of guanethidine in the rat and its uptake by heart slices. *Int. J. Neuropharmacol.* **4:**27–39.

Seabold, J. E., Cohen, E. L., Beierwaltes, W. H., Hinerman, D. L., Nishiyama, R. H., Bookstein, J. J., Ice, R. D., and Balachandran, S. (1976). Adrenal imaging with ^{131}I-19-Iodocholesterol in the diagnostic evaluation of patients with aldosteronism. *J. Clin. Endoc. Metab.* **42:**41–51.

Valk, T. W., Frager, M. S., Gross, M. D., Sisson, J. C., Wieland, D. M., Swanson, D. P., Mangner, T. J., and Beierwaltes, W. H. (1981). Spectrum of pheochromocytoma in multiple endocrine neoplasia. *Ann. Int. Med.* **94:**762–767.

Wieland, D. M., Swanson, D. P., Brown, L. E., and Beierwaltes, W. H. (1979). Imaging the adrenal medulla with an I-131-labeled adrenergic agent. Preliminary Note. *J. Nucl. Med.* **20:**155–158.

Wieland, D. M., Wu, J.-L., Brown, L. E., Mangner, T. J., Swanson, D. P., and Beierwaltes, W. H. (1980). Radiolabeled adrenergic neuron-blocking agents: Adrenomedullary imaging with [^{131}I]iodobenzylguanidine. *J. Nucl. Med.* **21:**349–353.

Wieland, D. M., Brown, L. E., Rogers, W. L., Worthington, K. C., Wu, J.-L., Clinthorne, N. H., Otto, C. A., Swanson, D. P., and Beierwaltes, W. H. (1981a). Myocardial imaging with a radioiodinated norepinephrine storage analog. *J. Nucl. Med.* **22:**22–31.

Wieland, D. M., Brown, L. E., Tobes, M. C., Rogers, W. L., Marsh, D. D., Mangner, T. J., Swanson, D. P., and Beierwaltes, W. H. (1981b). Imaging the primate adrenal medulla with [^{125}I] and [^{131}I] meta-iodobenzylguanidine: concise communication. *J. Nucl. Med.* **22:**358–364.

Wince, L. C., and Rutledge, C. O. (1982). Developmental changes in the accumulation of norepinephrine and the chronotropic effects of catecholamines in isolated rat atria as influenced by dietary lipid. *J. Pharmacol. Exp. Ther.* **220:**340–346.

CHAPTER 3

Monoclonal Antibodies: The Design of Appropriate Carrier and Evaluation Systems

BRIAN M. GALLAGHER

INTRODUCTION

The enthusiasm and appeal of using specific antibodies for the *in vivo* diagnosis and therapy of cancer over the last several years has shown ever-increasing growth. However, the widespread clinical application of these reagents has not kept pace with the promise. This has been due in part to the lack of systematic investigations of the many parameters that influence tumor localization. Too little attention has been paid to the appropriate requirements that the carrier antibody should possess, the antigen(s) that have served as the target, and the experimental animal models chosen for the preclinical evaluations.

The definition of specific tumor antigenicity was recognized by Witebsky in 1929, when he developed a rabbit antiserum against a uterine carcinoma without demonstrable cross-reactivity with normal tissue (Witebsky, 1929). This immediately suggested the possibility of serological therapy in cancer patients. The application of this knowledge to the experimental demonstration of external tumor localization did not occur until several years later (Pressman and Korngold, 1953; Pressman et al., 1957; Day et al., 1959). This early work was subsequently extended to human tumors by the use of affinity-purified antibodies to carcinoembryonic antigen (CEA) that were shown to localize preferentially in human colon tumors expressing CEA (Primus et al., 1973; Mach et al., 1974). Subsequently, numerous studies have extended these results to CEA (Goldenberg et al., 1978; Mach et al., 1980), human chorionic gonadotrophin (hCG) (Bagshawe et al., 1980), and alpha fetoprotein (AFP) (Goldenberg et al., 1980). Many of these studies have been carried out using conventional, polyclonal antibodies that required the use of significant background subtraction techniques to eliminate blood and nontumor-associated radioactivity.

The antibody response to a typical antigen is highly heterogeneous. When a foreign material is first exposed to the body's immune system, there is a rapid proliferation of lymphoid cells that can synthesize immunoglobulins that will bind to the invading antigen. There are as many as a million different lines of B lymphocytes in the spleen and the precursors of plasma cells that secrete the immunoglobulins. All are derived from a common stem cell, but each line develops an independent capacity to make an antibody that recognizes a different antigenic determinant. Usually, a number of different immoglobulins, each derived from a different cell clone (arising from a common parent cell), are produced during the primary response phase. This results in a heterogeneous population of immunoglobulins, each of which is specific for one of the chemical configurations found in the original antigen. Although all of these plasma cells are derived from a common stem cell, each clone line that develops in response to an antigen develops an independent capacity to synthesize antibody that specifically recognizes a different antigenic determinant. Thus the primary response includes diverse antibodies directed against different antigenic sites on the foreign antigen, different sites on a single determinant (epitopes), and even different antibodies with varying degrees of recognition and affinity to a single determinant. Therefore, conventional antisera and polyclonal antibodies are complicated, heterogeneous mixtures that may vary widely from animal to animal.

In 1975 Köhler and Milstein (1975) reported that somatic cell hybridization techniques could be used to establish continuous cultures of specific antibody-forming cells. The antibodies produced by such cultures are fundamentally different from antisera obtained from conventionally immunized animals. Each hybrid cell line or hybridoma produces a homogeneous or monoclonal immunoglobulin that represents only one of the many antibodies produced during the immune response of the host animal. This is due to the nature of the hybridoma technology and the clonal selection hypothesis advanced by Burnet in the 1950s (Burnet, 1959). The hypothesis states that each antibody-forming cell (B lymphocyte or its fully differentiated progeny, the plasma cell) is committed to the production of one type of antibody molecule that has the potential of reacting with one or at most a few structurally similar antigenic determinants. During the immune response those B cells recognizing a particular antigen in association with other immunoregulatory cells are stimulated to replicate and differentiate into antibody-secreting cells. Since most macromolecules have a number of distinct antigenic determinants, they stimulate the expansion of many B-cell clones, each secreting specific immunoglobulin molecules, specific for a different part of antigen (Yelton and Scharff, 1981).

Unfortunately, although techniques are currently available for separating the antibody-producing cells from an animal's spleen, lymph nodes, or bone marrow, the technology for culturing and proliferating clones from single cells is still under development. They therefore cannot be maintained in tissue culture for significant lengths of time to produce sufficient quantities of monoclonal antibody.

However, there are malignant tumors of the immune system, called myelomas that rapidly proliferate and produce large quantities of abnormal immunoglobulins, called myeloma proteins (Milstein, 1980). These cells can be cultured indefinitely, and all the immunoglobulins they secrete are identical in chemical structure. In 1975 Köhler and Milstein (1975) learned how to fuse mouse myeloma cells with lymphocytes from the spleen of mice immunized with sheep red blood cells (Fig. 3-1). The resulting hybrid-myeloma or hybridoma cells expressed both the lymphocytes' property of specific antibody production and the immortal character of the myeloma cell and could thus be maintained in tissue culture. Individual hybrid cells can be cloned and each clone can produce large amounts of identical antibody to a single

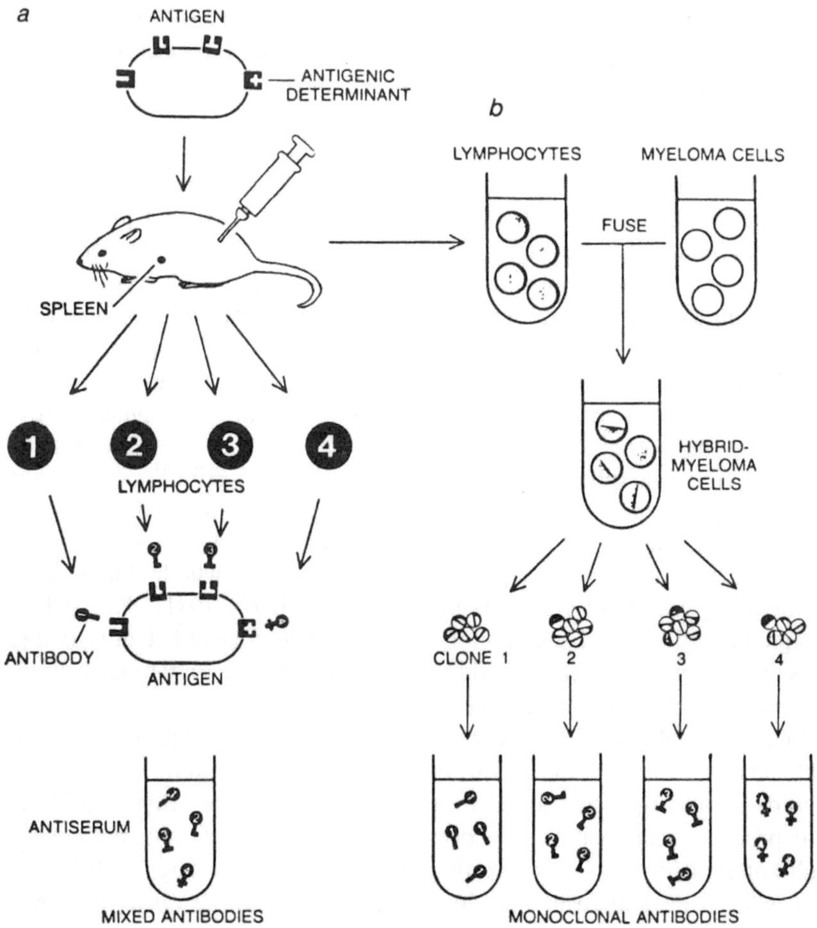

FIG. 3-1. Comparison of polyclonal antisera versus monoclonal antibodies.

antigenic determinant. These hybridomas can be injected into mice, and the resulting tumor grows in ascitic form, secreting tens of milligrams of antibody.

The production of hybridomas is now quite straightforward, although the technique is laborious and requires some skill (Fig. 3-2). Usually, mice or rats are immunized and boosted. The optimal immunization regimen will vary for each immunogen (substances capable of eliciting antibody responses). Of significance is the fact that unlike the production of conventional antisera or polyclonal antibodies, the immunogens used can be very impure, highly heterogeneous mixtures, such as tumor cell membranes, that may express thousands to hundreds of thousands of epitopes. With conventional sera, however, it is usually not possible to remove the most desirable and homogeneous antibody from the mixture present in the host animal's serum. With the application of the monoclonal approach, one screens the antibodies produced by each hybridoma clone and selects those that display the desired characteristics. It is therefore not necessary to have a homogeneous immunogen, which greatly facilitates the application of this method to the investigation of complex biological structures such as cell surface antigens.

At the time when the greatest numbers of antibody-forming cells can be found in the spleen, a single cell suspension of lymphocytes is mixed with myeloma cells, treated with polyethyleneglycol (PEG), and plated into a selective medium (Yelton and Scharff, 1981). This fusion protocol results initially in a mixture of unfused lymphocytes and myeloma cells, hybrids of each parent to itself and hybrids between a lymphocyte and myeloma cell. It is necessary to grow the cells in a medium that will only support the growth of these latter, hybridoma cells. This usually involves the use of hypoxanthine-aminopterin-thymidine (HAT) selection system. The myeloma fusion parent has been engineered with a genetic defect in the enzyme hypoxanthine-guanine phosphoribosyl transferase (HPRT). $HPRT^+$ cells cannot grow in the presence of 6-thioguanine, which provides a means to select for $HPRT^-$ cells. The $HPRT^-$ cells cannot use exogenous hypoxanthine to synthesize purines and die in the presence of aminopterin, which blocks *de novo* purine and pyrimidine synthesis (Yelton and Scharff, 1981). This defect is complemented by the B-cell genome in the hybridoma cell. Unfused spleen cells will not grow in culture so that after a few days the only rapidly dividing cells remaining are the myeloma–spleen hybrids.

In a few weeks these hybrid cells continue to grow and produce antibody that can be assayed by a wide variety of techniques (Milstein, 1980). Positive hybrids should then be cloned to avoid overgrowth by other hybrid cells. This process is most commonly done by making serial dilutions of cells into multicompartment vessels with the goal of obtaining a single cell in a well that will divide and form the clone line, producing the desired antibody. Growing hybrid cells on solid agar is another technique used. These subclones need to be rescreened to pick those secreting monoclonal antibody

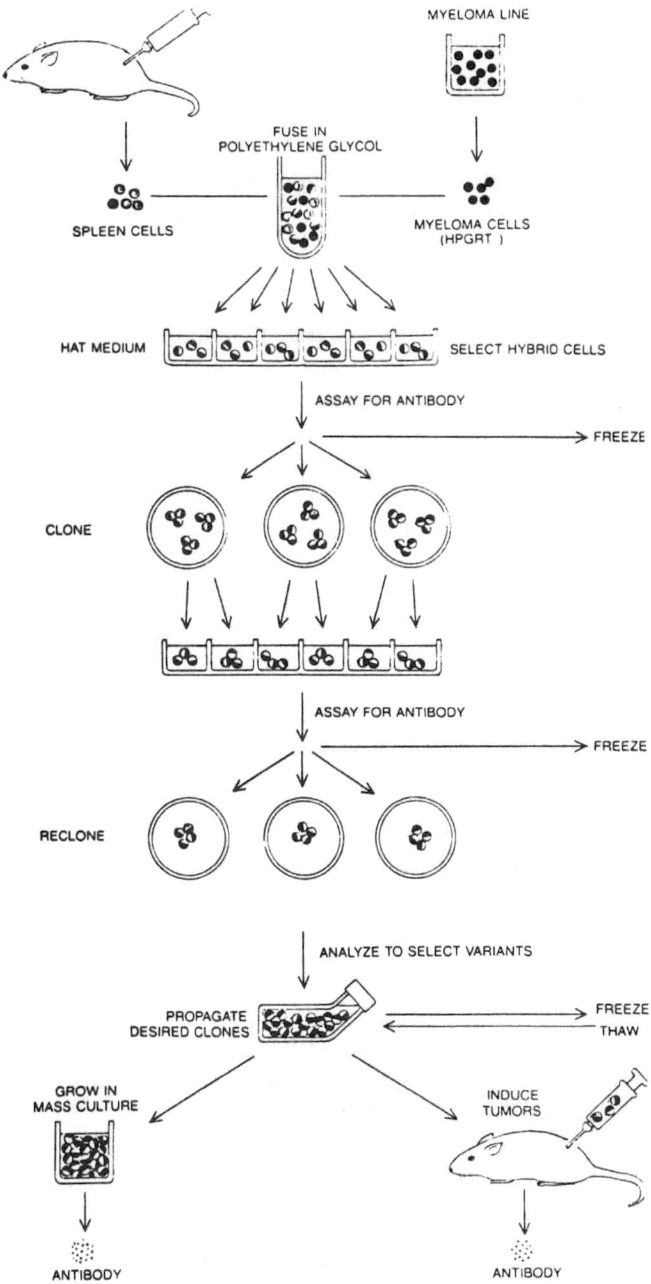

FIG. 3-2. Typical procedure for producing monoclonal antibodies.

with the desired specificity or predetermined requirements. When these are identified, they can be frozen for long-term storage and/or injected into animals of the same strain as those of the original donor cells used for the fusion to produce ascites fluid containing the desired immunoglobulin. Alternatively, the clone can be grown in mass tissue culture and the antibody harvested from the supernatant; however, the yields by these tissue culture methods are significantly lower than those by the ascites route.

MONOCLONAL ANTIBODY APPLICATIONS

Over the last few years, a number of potential applications of this powerful technology have been suggested. It is appropriate to reemphasize the word *potential,* since, in many cases, the demonstration of utility or efficacy has not yet been made. Table 3-1 lists some of the more commonly suggested uses of monoclonal antibodies. The use of monoclonal antibodies for passive immunization against infectious diseases and autoimmune disorders, against graft–transplantation rejection, and as therapeutic delivery systems *in vivo* will require improvements in the current technology. It would be preferable to use human–human hybridomas, formed by the fusion of a human antibody-secreting cell and a suitable human fusion partner to immortalize the line, for these repeated *in vivo* applications. This would avoid stimulation of the host human immune system to form a response to the repeated challenge with foreign (mouse) proteins. The possibility of this approach has been recently demonstrated (Olsson and Kaplan, 1980), and many improvements in this technology are under intense investigation. Other efforts, still very much in the research phase, include attempts to maintain antibody-producing human B cells in continuous culture using the recently described immunoregulation factors, the viral transformation of human B cells to immor-

TABLE 3-1. Monoclonal Antibody Applications

1. Passive immunization
   a. Infectious diseases
   b. Autoimmune disorders
   c. Drug toxicity
2. Graft–Transplant rejection
3. Manipulation of immune response
4. *In vitro* specific assays
5. Affinity purification methods
6. Cell surface mapping
7. Delivery systems
   a. Liposomes
   b. Cytotoxic agents
   c. Detection of abnormalities

talize them, and human immunoglobulin production using recombinant genetic techniques.

There are a number of persisting technical problems with respect to the application of this technology. These are outlined in Table 3-2. Some immunogens are very limited in the ability to elicit a strong response in the host animal. There are a variety of ways in which to stimulate the immune response, but it is frequently very difficult and labor intensive to screen large numbers of hybrid clones to look for the relatively small number of lines secreting the antibodies with the desired characteristics.

Unlike polyclonal antibodies, which may represent hundreds of different proteins, each unique in structure but all perhaps with recognition for different epitopes on a single antigen molecule, a monoclonal antibody is a distinct, unique protein with a single amino acid composition, sequence, and structure. In many cases, a particular monoclonal protein may be more or less susceptible to inactivation, denaturation, or changes in specificity caused by pH, ionic strength, labeling with radionuclides, etc. If the particular protein with which one is working displays these sensitivities, a marked loss in immunoreactivity and biological activity will be evident. With polyclonal antibodies, one can sustain loss of many of the more sensitive immunoglobulins in the mixture with the preservation of much of immunoreactivity to a particular antigen by the more resistant proteins. What one loses in the latter case is the recognition of specific epitopes on the antigen, but other immunoglobulins in the mixture will still react with other sites on the same antigen.

At present, the species most commonly employed for making monoclonal antibodies include mouse and rat, mostly because the myeloma lines in culture that fuse well are from these species (Yelton and Scharff, 1981). Hybrids between mouse myeloma and rabbit lymphocytes segregate the rabbit chromosomes extremely rapidly and stop producing rabbit immunoglobulin. Hybrids formed between mouse myeloma cells and human lymphocytes are somewhat more stable and a number of hybrids secreting human immunoglobulin have been generated (Levy and Dilley, 1978). In

TABLE 3-2. Persisting Technical Problems

1. Weak immunogens
2. Sensitivity to inactivation
3. Species limitations
4. Labor intensive
5. Validity of screening procedures
6. Purification on large scale
7. Large-scale production
8. Need for panel of antibodies
9. Protein modification

general, then, studies on the utility of these agents in animal models will be greatly limited to studies that can be carried out in mice and rats. The implications of this and how it affects the choice and design of animal models will be discussed later in this chapter.

At present, the large-scale production of hundreds of milligrams of antibody is limited to ascites methods in rats and mice. Although the ascites is greatly enriched in the specific monoclonal antibody being secreted, ascites fluid contains many other proteins, including albumin, transferrin, and nonspecific immunoglobulins. These impurities must be removed by suitable chromatographic or electrophoretic methods to avoid potential problems with labeling with radioisotopes.

Because of the precise specificity that one can obtain with a monoclonal antibody to a particular site on a specific antigen, the number and availability of such sites may determine the total amount of antibody that will bind *in vivo*. It will likely be necessary, therefore, to produce panels of monoclonal antibodies with different epitopic recognition for a particular cell surface antigen to obtain the concentration of antibody at the target site sufficient for external detection. Although this suggests a similarity with a polyclonal serum, it differs in that the panel will be in the tens, not hundreds, of antibodies. Each component of the panel will be of defined structure and behavior, and uniform material can be produced indefinitely, unlike polyclonal serum, which varies from animal to animal and where availability stops when the donor animal dies.

An area of particular interest to the radiopharmaceutical scientist is in the development of new and better methods for protein modification and labeling methods. Although a number of methods have been developed (Pettit et al., 1980; Krejcarek and Tucker, 1977; Khaw et al., 1980; Paik et al., 1980; Eckelman et al., 1980; Paik et al., 1982; Van Nagell et al., 1980), there is a tremendous need for methods that preserve both the immunoreactivity (recognition of the specific antigen) and biological activity (clearance and metabolism *in vivo*). Many of these technical issues have not been addressed adequately in conjunction with labeling studies, but the successful application of these techniques will be heavily dependent on the rigors of such determinations. It is not the purpose of this chapter to review the current state of the art of this area, but it represents an extremely interesting and important aspect of the utility of monoclonal antibodies.

REQUIREMENTS FOR THE RECEPTOR ANTIGEN

Table 3-3 lists some of the more important characteristics that are desirable in the production of monoclonal antibodies against tumor-associated antigens. Some controversy exists at present on the effect of circulating free antigen on the imaging properties of labeled antibodies. Successful imaging studies have been carried out in patients with high circulating levels of CEA (Van Nagell et al., 1980). These results are rather surprising, as it is gener-

TABLE 3-3. Requirements for the Receptor Antigen

1.	Nonshedding
2.	Nonmodulating
3.	Present on variants
4.	High density of expression
5.	Tumor specific or associated
6.	Expression independent of cell cycle

ally considered undesirable to inject antibodies against shed antigens *in vivo* both because of competition for radiolabeled antibody for the target site and because of the likely formation of immune complexes that, in sufficiently high concentrations, can cause kidney damage.

In addition, it is desirable that the cell surface antigen against which monoclonal antibodies are made does not modulate. Antigen modulation is a phenomenon whereupon following attachment of free antibody to a cell surface molecule, the entire complex is internalized or shed by one or more mechanisms. This behavior is not unusual, particularly in leukemias. The results of such internalization is a more rapid turnover and possible loss of radionuclide from the target site. Although this result would not be favorable for imaging tumors, it may be desirable for drug delivery purposes.

Perhaps one of the most important requirements for tumor-associated antigens is their presence on metastatic cells, which may themselves represent variants from the primary tumor line. The lethality of most cancers can be attributed to the ability of tumor cells to spread to and proliferate in organs distant from the primary site of origin (Hart and Fidler, 1981; Fidler and Hart, 1981). Metastasis is the formation of variants that no longer have continuity with the primary tumor. These changes may involve the acquisition or loss of various phenotypic properties, including immunogenicity (Killion and Kollmorgen, 1976; Kerbel, 1979) and radiation sensitivity (Fidler, 1978). This phenomenon suggests the probable need for panels of monoclonal antibodies directed against a number of tumor-associated antigens so that antigen-negative mutant cells can be detected.

It is also desirable that these surface antigens be present in high density and be preferentially expressed on tumor cells compared with normal tissues. These latter requirements are particularly of importance for imaging purposes if one is to achieve sufficient target-to-nontarget ratios of radioactivity to permit *in vivo* detection. The whole issue of "tumor specificity" for antigens is one that is far from resolved, but antigens that are preferentially expressed on abnormal cells, compared with normal, should be the goal. A host of publications with such claims have appeared in the last few years, a few of which are cited here to illustrate the literature (Carey et al., 1976; Yeh et al., 1979; Colcher et al., 1981; Mitchell et al., 1980; Crawford et al., 1981; Tax and Mason, 1981; Cuttitta et al., 1981; Old, 1981; Sibora and Wright, 1981; Kasai et al., 1981; Embleton et al., 1981; Ashall et al., 1982; McGee et al., 1982). These and other studies have led to a rough classification of surface antigens into three classes (Old, 1981).

Class 1 antigens are restricted to autologous tumor and cannot be found on autologous normal cells or any other normal or malignant cell type. Class 2 antigens are shared tumor antigens, found on autologous as well as allogeneic tumors of similar and dissimilar origins. They are not present on normal B cells, kidneys, or fibroblasts. Class 3 antigens are widely distributed on normal and malignant cells and are therefore normal cell surface antigens. Although intuitively, the Class 1 antigens would represent truly tumor-specific antigens and, therefore, the preferable target to produce monoclonal antibodies against, their existence is in question. Class 2 antigens represent the target to which most workers have produced monoclonal antibodies. It is also important that the expression of a given antigen on the cell surface be independent of the phase of the cell cycle, in order to ensure its continued presence in a heterogeneous cell population.

REQUIREMENTS FOR THE CARRIER ANTIBODY

Some of the requirements desirable for the carrier antibody are listed in Table 3-4. One feature of monoclonal antibody technology that has not always been observed is the production of homogeneous antibody. By homogeneous antibody, it is meant that both the heavy and light chains synthesized by the hybridoma line are derived from the genetic information of the host B cell, with no expression of the nonspecific myeloma fusion parent light or heavy chains. This is illustrated in Fig. 3-3. Fortunately, hybrid cells, particularly in the early stages of proliferation, tend to lose chromosomes rapidly. It is therefore possible to select variants that have lost the ability to synthesize nonspecific heavy (gamma) and light (kappa) chains and produce only the true, homogeneous monoclonal antibody coded for by the host spleen cell. This avoids the problems associated with mixed immunoglobulin molecules such as decreased amounts of the desired monoclonal produced and altered antigen specificity. Table 3-5 lists some of the cell lines that are

TABLE 3-4. Requirements for the Carrier Antibody

1. Homogeneity (only specific light chains)
2. Specificity (no cross reactivity)
3. Affinity (highest possible)
4. Isotype (effector functions)
5. Purity (ascites impurities)
6. Stability (IgM unstable; chemical modification)
7. Surface groups available (for coupling)
8. Non-antigenic (fragments; human antibodies)

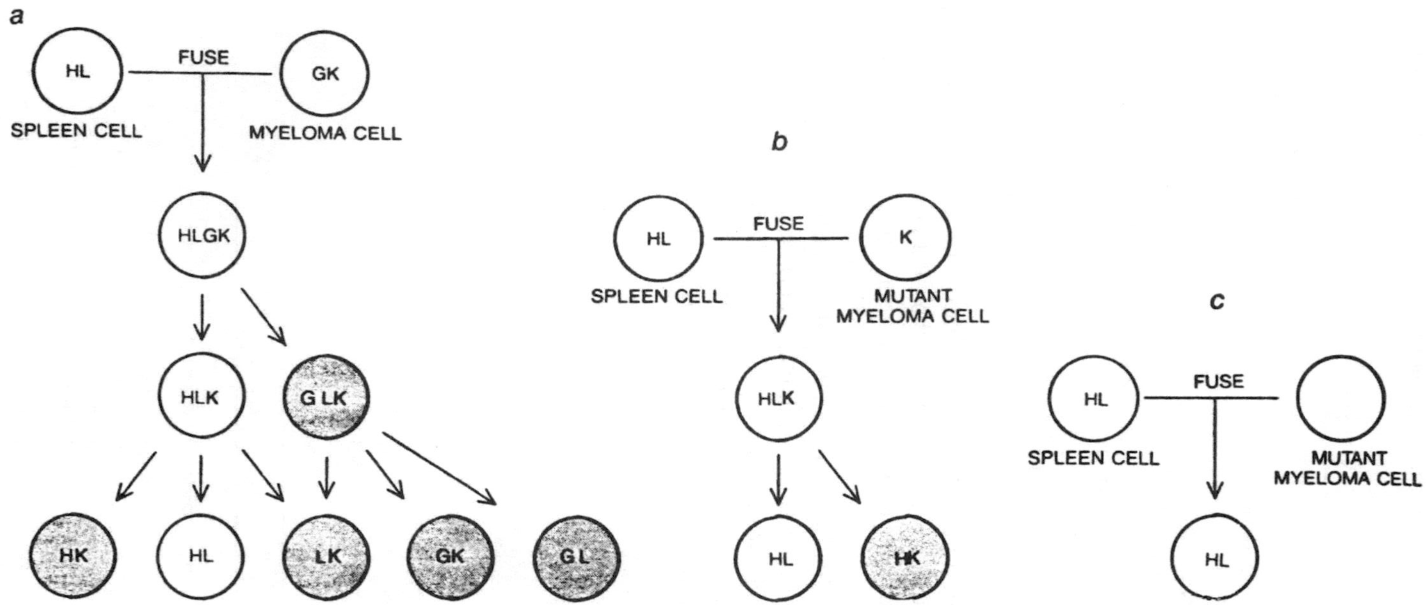

FIG. 3-3. Possible heterogeneity of secreting myeloma fusion parents.

TABLE 3-5. Common Myeloma Parents

Cell Line	*Immunoglobulin Produced*
P3-X63Ag8	$\gamma 1$, κ
45.6TG1.7	$\gamma 2b$, κ
NSI/1-Ag4.1	κ (Not secreted)
P3X63Ag8U$_1$	κ (Not secreted)
Sp2/0-Ag14	None
X63-Ag8.653	None
Y3-Ag1.2.3 (rat myeloma)	κ (Rat)

commonly used to generate hybridomas. The availability of nonproducing fusion parents makes it desirable to choose a nonsecreting fusion line from which to grow hybridomas that will then secrete only homogeneous immunoglobulin.

Figure 3-4 illustrates just how nonhomogeneous a "monoclonal" antibody can be produced by using inappropriate myeloma fusion partners. A hybridoma from a mouse spleen cell (mouse lymphocyte) that secretes only one heavy chain and one light has an amino acid sequence and hence structure that will determine the specific antigenic recognition. If this cell is fused

	H^M	H^L	L^M	L^L
H^M	H^MH^M	H^MH^L	—	—
H^L	H^LH^M	H^LH^L	—	—
L^M	—	—	L^ML^M	L^ML^L
L^L			L^LL^M	L^LL^L

$H^MH^ML^ML^M$	nonspecific	$H^LH^ML^ML^M$	nonspecific
$H^MH^ML^ML^L$	nonspecific	$H^LH^ML^ML^L$	nonspecific
$H^MH^ML^LL^M$	nonspecific	$H^LH^ML^LL^M$	univalent
$H^MH^ML^LL^M$	nonspecific	$H^LH^ML^LL^L$	univalent
$H^MH^LL^ML^M$	nonspecific	$H^LH^LL^ML^M$	nonspecific
$H^MH^LL^ML^L$	univalent	$H^LH^LL^ML^L$	univalent
$H^MH^LL^LL^M$	nonspecific	$H^LH^LL^LL^M$	univalent
$H^MH^LL^LL^L$	univalent	$H^LH^LL^LL^L$	bivalent

FIG. 3-4. Example of heterogeneity from secreting fusion parents.

with a myeloma fusion partner that also secretes a nonspecific light and heavy chain, the subsequent array of products is illustrated. Only one-sixteenth of the intact immunoglobulin molecules produced will have the "correct" light and heavy chain combination which encodes for the original recognition by the mouse lymphocyte product ($H^L H^L L^L L^L$). Six-sixteenths of the products will have univalent recognitions (one of the two Fab arms will react specifically with the target antigen), although such univalent attachment usually results in a much lower overall avidity than bivalent attachment. The remaining nine-sixteenths of immunoglobulins produced will have little or no recognition of the original antigen, since each arm will have at least one nonspecific peptide chain. Thus the "monoclonal" product of such a fusion is quite a heterogeneous mixture both in terms of protein structure and, more important, in terms of immunospecificity. It would be predicted that if one were to radio-iodinate this mixture and if one assumes equal facility in labeling all the chains present, then the nonimmune globulins would contribute greatly to the overall body background, thereby decreasing the tumor to nontarget ratios significantly.

As highlighted earlier, it is important that the monoclonal antibody show a specificity for the tumor type being studied with little or no cross-reactivity with normal human tissues. Since localization is largely a process governed by equilibria into and out of various body compartments, it is desirable to produce antibodies of the highest possible affinity for the tumor cell surface. This requirement is similar to that required for receptor ligands to have utility as radiopharmaceuticals (Eckelman et al., 1979). This selection process for affinity can be made during the screening process to identify clones producing antibody with the desired characteristics. This can be done using conventional methods (Sips, 1948; Scatchard, 1949; Karush, 1957; Schwartz et al., 1978; Frankel and Gerhard, 1979; Eisen, 1980).

An important feature of the carrier antibody is the class and isotype to which the specific molecule belongs. Five classes of human immunoglobulins have been identified and are designated as IgG, IgA, IgM, IgD, and IgE. The basic structural unit of all immunoglobulin (Ig) molecules consists of two pairs of disulfide-linked polypeptide chains, the heavy and light chains (Fig. 3-5). One disulfide bond attaches each light chain to one of the heavy chains. The two heavy chains are connected to each other by one or more disulfide bonds. In addition to interchain bonds, both light and heavy chains have intrachain disulfide bonds. The IgG class can be subsequently further divided into IgG1, IgG2, IgG3, and IgG4 and the IgA into IgA1 and IgA2. Each of the nine total classes has antigenically distinctive heavy chains (named with the corresponding Greek letter—α, γ, μ, δ, and ε), but the light chains are similar in all cases (Eisen, 1980). Since about 85% of the Ig in human serum is of the IgG class, the bulk of the description of immunoglobulins will be restricted to these molecules, although the picture is considerably more complex than appropriate for this discussion (Eisen, 1980; Atassi, 1979).

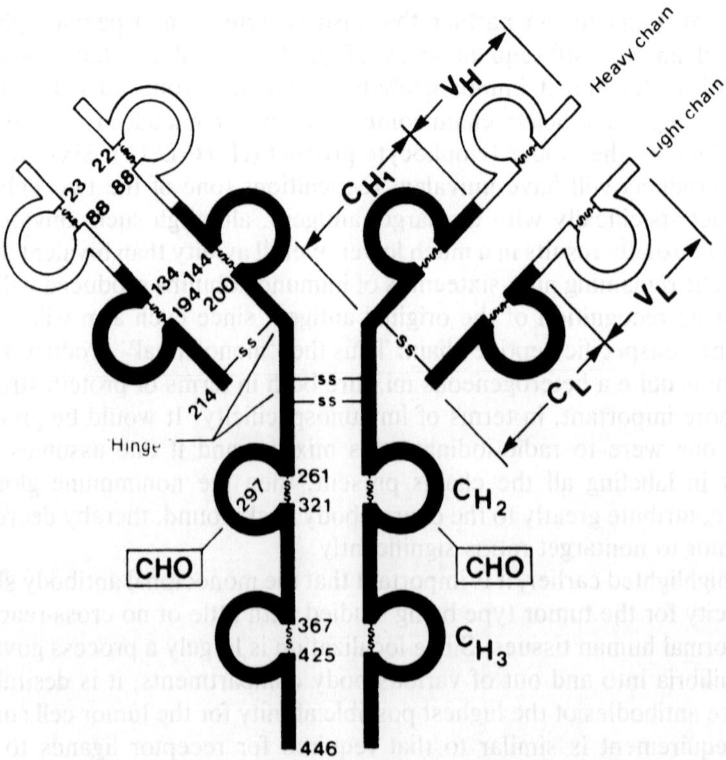

FIG. 3-5. Schematic diagram of IgG molecule.

The amino terminal portions of heavy and light chains are characterized by a variable amino acid sequence, designated V_H and V_L, respectively, whereas the carboxyl-terminal portions of each chain, C_H and C_L, have a relatively constant sequence (Fig. 3-6). The variable and constant regions of both chains are composed of a series of structurally compact, homologous domains, each containing about 110 amino acid residues. The V_H and V_L domains contribute to the antigen-binding site and determine the specificity, affinity, etc., for a particular antigen. The remaining C_H domains are involved with biological effector functions such as complement-binding, skin fixation, placental permeability, etc. (Putnam, 1977).

The proteins belonging to each class of Ig possess distinctive chemical, physical, genetic, metabolic, and other biological properties, all of which may exert a profound influence on their utility for *in vivo* diagnosis and therapy. A summary of some structural and functional properties of the immunoglobulins is shown in Table 3-6 (Soloman, 1980). The serum levels of Ig molecules are dependent on their rates of synthesis, catabolism, and distribution between intravascular and extravascular compartments (Waldman et al., 1970). This is illustrated in Table 3-7, and it is obvious that there are striking class-related differences among Ig's representative of each of the

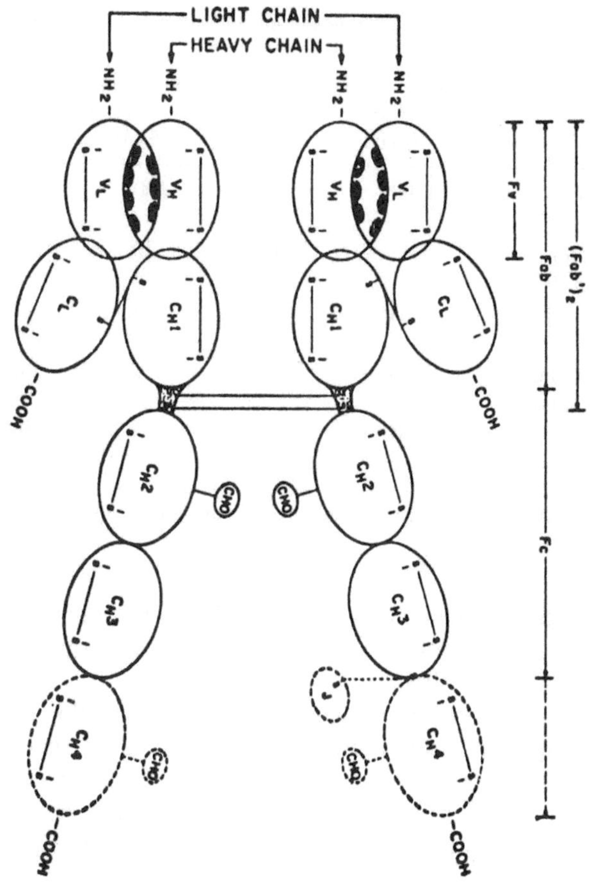

FIG. 3-6. Diagram of IgG showing location of domains.

TABLE 3-6. Structural and Functional Properties of Human Immunoglobulins

Property	*IgG*	*IgA*	*IgM*	*IgD*	*IgE*
Molecular weight	160,000	170,000	900,000	180,000	200,000
$S_{20,w}$	7	7a	19[a]	7	8
J chain	0	0[b]	+	0	0
Carbohydrate, %	2.9	7.5	10.7	12.0	10.7
Genetic factors	Gm	Am[c]	?	?	?
Complement	C[d]	A[e]	C	A	A

[a] The IgA molecule is usually present in serum as a 7 S monomer but also occurs as 9, 11, and 13 S polymers. Higher polymeric forms of IgM have been detected in macroglobulinemic serums; the monomeric (7 S) form of IgM may also be found (Solomon, 1980).

[b] Only one J chain has been found associated with any of the polymeric forms of IgA and IgM. The monomer forms of IgA and IgM are devoid of J chain.

[c] Am factors have been detected only in proteins of the IgA2 subclass.

[d] Classical pathways.

[e] Alternate pathways.

TABLE 3-7. Metabolic Properties of Human Immunoglobulins

Property	*IgG*	*IgA*	*AgM*	*IgD*	*IgE*
Serum concentration (mg/dl)	1200	250	100	2.3	0.05
Intravascular (IV) concentration (%)	45	42	76	75	51
Biological survival $t_{1/2}$ (days)	23[a]	5.8	5.1	2.2	2.4
Fractional catabolic rate (% IV pool/day)	0.07	0.25	0.19	0.37	0.89
Synthetic rate (mg/kg day)	34	24	7.9	0.40	0.02

[a] For IgG3 proteins the $t_{1/2}$ synthetic and catabolic rates are 7 days, 3.4 mg/kg day, and 16.8%, respectively (Solomon, 1980).

five established heavy chain classes. Furthermore, the catabolic rate of IgG is concentration dependent—the higher the serum concentration, the faster its catabolism (Soloman, 1980).

Antigenic analysis with heterologous antisera revealed that proteins of the human IgG group can be differentiated into four classes (IgG1–IgG4) each with distinctive heavy chain (γ 1 through γ 4). IgG1,2,3,4 make up about 70%, 19%, 8%, and 3%, respectively, of human IgG proteins (Eisen, 1980). A stretch of about 15 amino acid residues between C_{H_1} and C_{H_2} corresponds

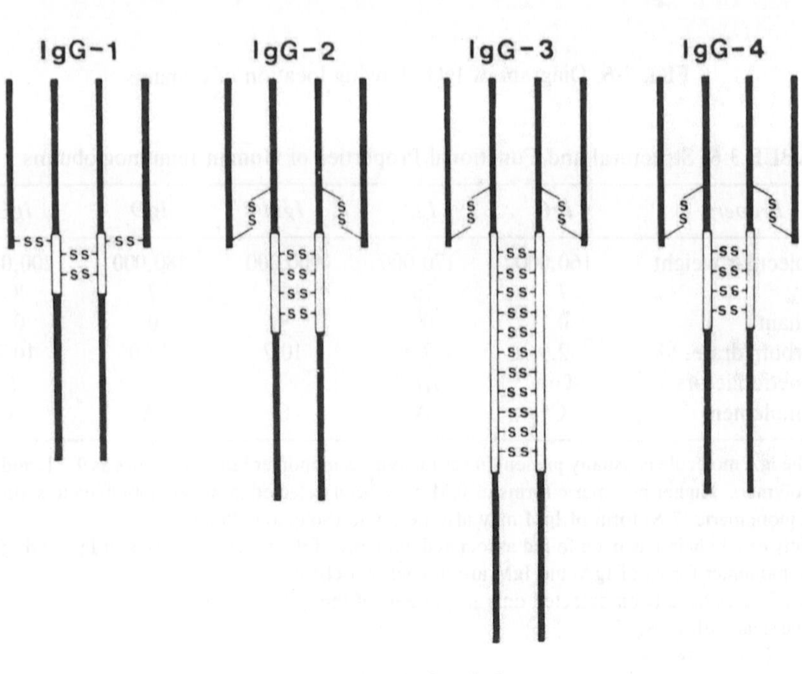

FIG. 3-7. The IgG subclasses.

to the flexible junction between the so-called Fab (fragment antigen binding) and Fc (fragment crystallizable) portions of the molecule in $\gamma 1$, $\gamma 2$, and $\gamma 4$ chains (Fig. 3-7). This so-called hinge region has no obvious homology to other domains and contains the disulfide bonds that link the two heavy chains of each molecule and the $\gamma 1$ and light chains of IgG1. The hinge is about four times longer and has about four times more disulfide bonds between the heavy chains in $\gamma 3$ than in the other γ chains, owing to duplications of the normal hinge sequence.

The few cleavages produced in IgG's in response to proteolytic enzymes in native immunoglobulins are localized in the hinge region (Fig. 3-8). The abundance of proline and cysteine residues in this region apparently confers rigidity to this portion of the molecule, which exposes this site to proteolytic enzymes (Nisonoff et al., 1975; Anzel and Poljak, 1979). Cleavage with papain yields fragments that retain the specific combining sites of the parent molecule of ~50,000 mol wt but are univalent, i.e., they can bind one antigen per fragment and constitute the Fab fragments (Porter, 1959). Pepsin also

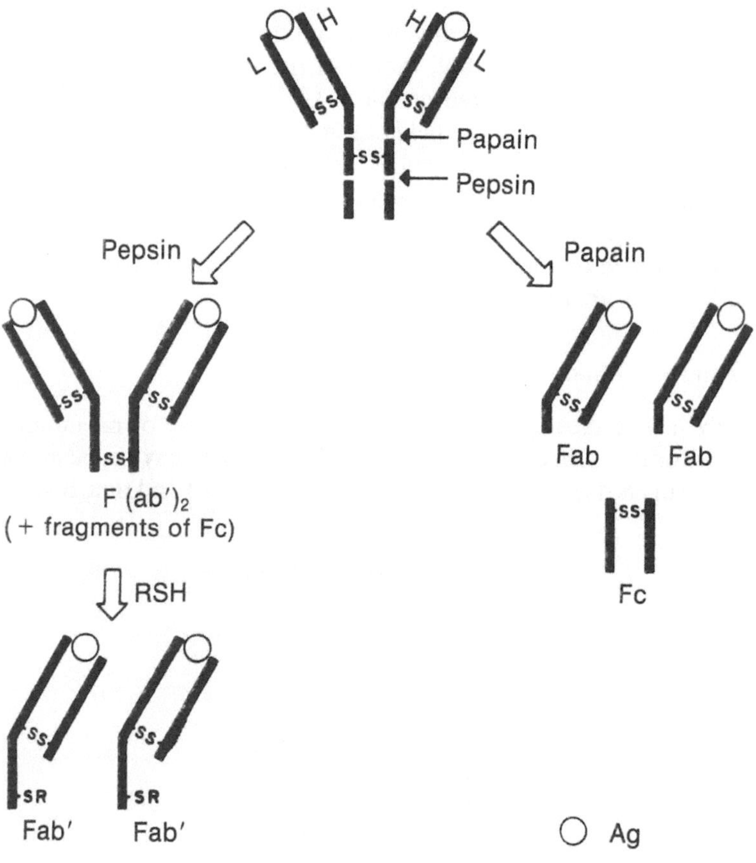

FIG. 3-8. Pepsin–papain cleavage of antibodies into fragments.

cleaves IgG antibodies but yields fragments of ~100,000 mol wt that are bivalent with regard to binding of antigens (as are the parent molecules). These fragments are termed $F(ab')_2$ fragments and can be split subsequently by reduction of the disulfide bonds between the heavy chains to yield Fab' fragments. These are indistinguishable from the Fab fragments with respect to antigen binding but have about a 10% greater molecular weight due to a slightly longer heavy chain. These fragments have been shown to possess full immunoreactivity *in vivo* (Smith et al., 1971) and are excreted largely into the urine, which results in a biological half-life usually less than 5 hr for Fab fragments (Spiegelberg and Weigle, 1965; Waldmann and Strober, 1969; Butler et al., 1973; Smith et al., 1977; Smith et al., 1976; Khaw et al., 1976; Burchiel et al., 1982). In addition, the removal of the Fc portion of immunoglobulins greatly reduces their antigenicity in a xenogeneic host.

TUMOR MODEL SYSTEMS

With the adherence to the general guidelines for choosing the appropriate immunoglobulin with specificity against an appropriate antigen, the choice of a relevant animal model for radiolabeled antibody studies remains as perhaps the greatest potential pitfall. Some of the more commonly employed tumor model systems are listed in Table 3-8. It is not the purpose of the present discussion to review the field of tumor models for radiopharmaceutical research, as this has been the subject of Chapter 4. It is of importance to note that the most commonly employed tumor models for studies on the localization properties of antibodies are restricted largely to murine or rat tumors and human tumor xenografts growing either in athymic "nude" animals or immunosuppressed mice.

Murine and Rat Tumors

Relatively few studies on the use of radiolabeled mouse or rat monoclonal antibodies for the *in vivo* detection of allogeneic tumors have appeared in the literature. This is due, in part, to the considerable effort and time involved in

TABLE 3-8. Tumor Model Systems

1. Murine tumors
2. Nude mice
3. Nude rats
4. Immunosuppressed mice
5. Hamster cheek pouch
6. Rabbit cornea
7. *In vitro* models

the production of high affinity, highly specific, and pure monoclonal antibody that would have an ultimate utility limited to experimental animal studies. This is a most unfortunate circumstance in that the wide variety and diversity of spontaneous and transplantable rodent tumor models available for general, nonimmunologically based radiopharmaceutical research have not been explored to establish the feasibility of *in vivo* tumor detection. In addition, many of the studies reported in the literature are difficult to interpret because of the antibodies, the animal model, or the experimental design chosen for the evaluation.

The external detection of a murine teratocarcinoma using a radiolabeled monoclonal antibody has been reported (Ballou et al., 1979; Levine et al., 1980). Figure 3-9 shows an image obtained five days after the injection of [^{131}I]antibody showing localization in the NIH-15 teratocarcinoma and no visualization of a nonspecific tumor. Unfortunately, the extremely large size of the tumor is not relevant to the use of this approach in imaging regional

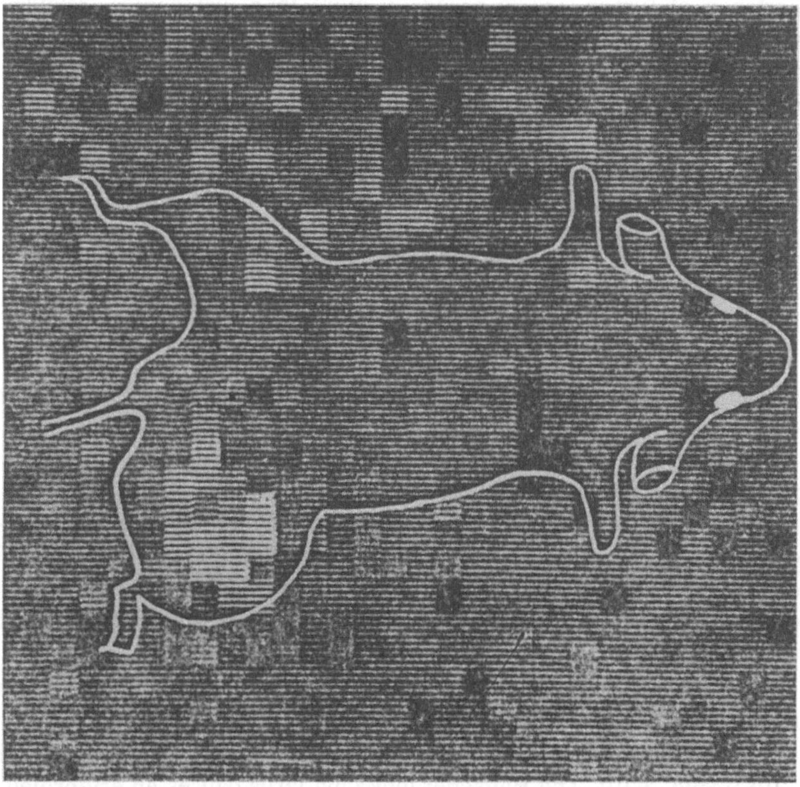

FIG. 3-9. Image of a mouse bearing an MH-15 teratocarcinoma in the left side and a nonspecific P3-X63-Ag8 myeloma tumor in the right side, 5 days postinjection of monoclonal antibody. (From Ballou et al., 1979).

and distal metastases and small primary tumors. Although tumor weights are not given, the weight of a spherical tumor with a 25-mm diameter can be approximated using the NCI guidelines as follows (Geran et al., 1972):

$$\text{Tumor weight (mg)} = \frac{\text{length (mm)} \times (\text{width [mm]})^2}{2} \qquad \text{(Eq. 3-1)}$$

or in the previously cited example:

$$= \frac{(25) \times (625)}{2} = 7.8 \text{ g}$$

If one assumes an average body weight of 30 g for a mouse, a tumor of this size represents about 26% of the body weight. The human equivalent of this would be an 18.2-kg tumor in a 70-kg man or a 40-lb tumor in a 150-lb man. In fact, even a 1-g tumor in a "standard" mouse represents about 3.3% of the body weight, which would translate to a 2.3-kg or a 5-lb tumor in a human. This is approximately 1.5× the weight of the human brain and hardly represents the clinical needs for the early diagnosis of tumors by imaging. The importance of tumor size is based on more than just clinical relevance, however, as it has been recently shown that the specificity of tumor uptake can be influenced by tumor weight (Moshakis et al., 1981; Moshakis et al., 1981). Specificity in this case was defined in terms of a localization index expressing the ratio of an ^{125}I specific monoclonal antibody and an ^{131}I nonspecific monoclonal antibody that are coinjected. The localization index in the tumor and various organs is then calculated as follows:

$$\text{Localization index} = \frac{\dfrac{\text{tumor}^{125}\text{I activity/weight}}{\text{tumor}^{131}\text{I activity/weight}}}{\dfrac{\text{blood}^{125}\text{I activity/weight}}{\text{blood}^{131}\text{I activity/weight}}} \qquad \text{(Eq. 3-2)}$$

This method allows specific and nonspecific activity to be compared between tumors and tissues while taking into account blood-borne activity. Such studies have shown a consistent observation that the smaller the tumor, the higher the localization index (Moshakis et al., 1981; Moshakis et al., 1981). This observation could be due to many factors, such as increased necrosis in larger tumors, increased vascularity, lymphatic drainage, differences in extracellular fluid space, etc. As few studies have utilized this approach, the general applicability to a wide range of tumor types needs to be established.

Recent studies have appeared on the use of the Rauscher murine erythroleukemia model to investigate the targeting of radiolabeled monoclonal antibodies to neoplastic cells *in vivo* (Sheinberg and Strand, 1982; Sheinberg et al., 1982) (Fig. 3-10). The Rauscher leukemia virus causes an erythroleukemia characterized by rapid transformation of erythroid precursor cells, splenomegaly and viremia. After virus infection in mice, neoplastic erythro-

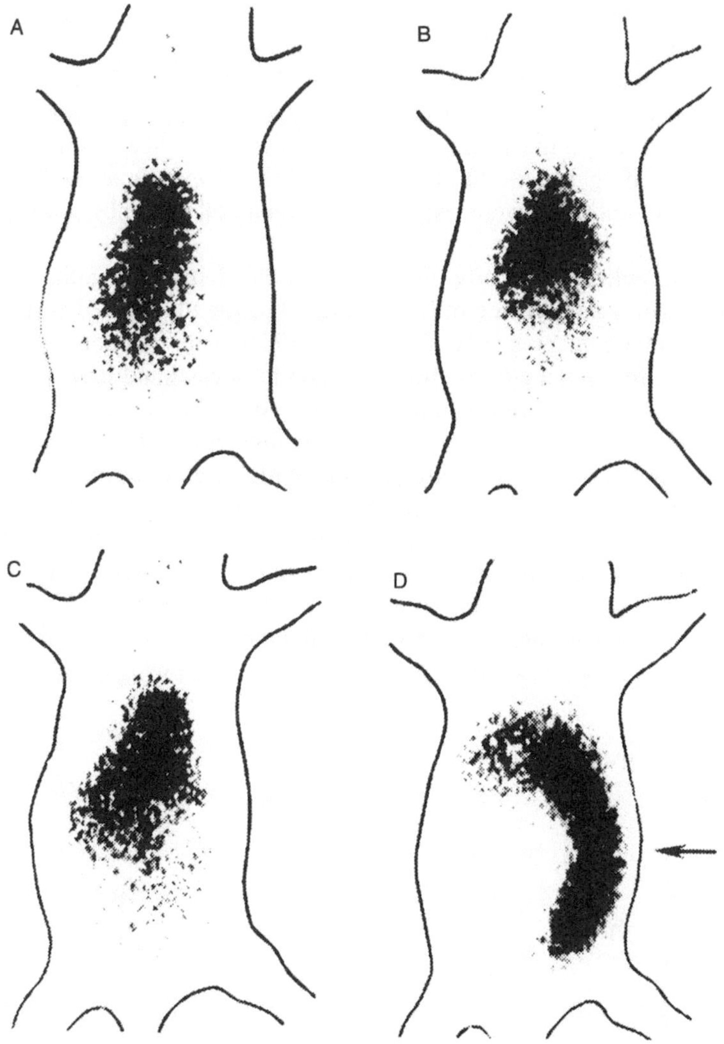

FIG. 3-10. Gamma camera images of normal and leukemic mice injected with [^{111}In] DTPA-103A and [^{111}In]DTPA-263D (a nonspecific control antibody at 5–6 hr postinjection). Total counts amounting to 4×10^4 were accumulated for each image with the animals positioned on their backs. (A) Normal mouse with [^{111}In]263D, (B) normal mouse with [^{111}In]103A, (C) leukemic mouse with [^{111}In]263D, (D) leukemic mouse with [^{111}In]103A. The leukemic spleen (arrow) contained 15% of the injected dose of [^{111}In]103A. Spleen weights of the four mice were, respectively, 118, 98, 652, and 339 mg. (From Sheinberg *et al.*, 1982.)

blasts in the spleen are seen microscopically within a few days, and macroscopic tumor foci are visible after about one week; finally, about a 10-fold increase in spleen size may occur in about 2 weeks (Sheinberg and Strand, 1982; Rausher, 1962). The virus envelope glycoprotein, GP 70, is expressed in large amounts on the surface of virus-infected cells. Monoclonal antibody 103A directed against these cells indicated that approximately 10^6 binding sites for this antibody were expressed on each cell (Sheinberg and Strand, 1982).

The monoclonal antibody 103A was modified by conjugation with the carboxycarbonic anhydride of DTPA and subsequently labeled with ^{111}In for leukemic spleen imaging (Sheinberg et al., 1982). The ^{111}In conjugated IgG's were tested for specificity by direct binding to normal and leukemic cells *in vitro* (Fig. 3-11). The monoclonal 103A bound to the leukemic spleen cells but not to normal spleen cells. Control immunoglobulins that were isotype matched IgG1's but had no antigenic recognition for either normal or leukemic spleens showed no significant binding, as expected. It should be noted, however, that only about 10% of the amount of ^{111}In-labeled specific antibody added to the cell culture actually bound to the target cells. Thus although the specificity of binding was maintained, the degree of binding was quite low. This illustrates a most important factor that should be investigated in all studies employing radiolabeled antibodies. Following chemical modifi-

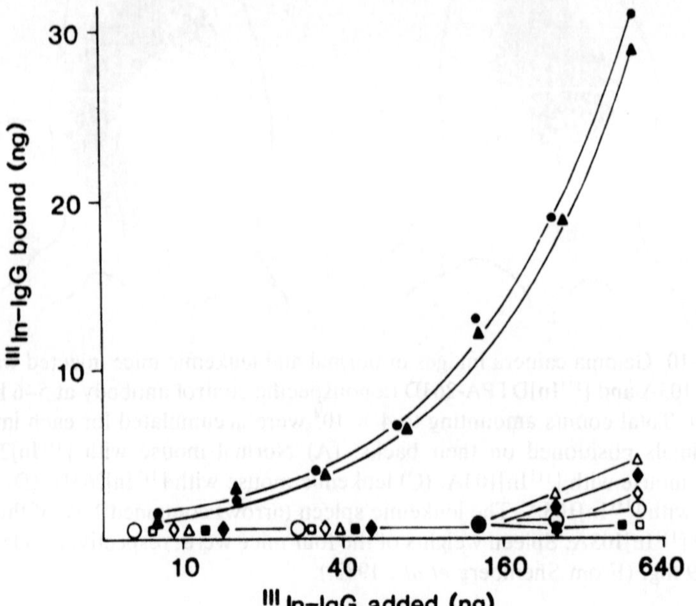

FIG. 3-11. Direct binding assay of live normal (○, □, △, ◇) and leukemic (●, ■, ▲, ◆) spleen cells to IgG's conjugated with [^{111}In]DTPA and [^{111}In]EDTA measured *in vitro*. (From Sheinberg *et al*., 1982.)

cation and labeling, the specificity of the antigen recognition on the target cells should be measured relative to a nontarget cell and relative to isotype-matched control immunoglobulin. This should measure not just the qualitative difference in the binding parameter but also the maximum percent of labeled antibody bound under antigen excess conditions.

The maximum binding can be measured either by increasing cell number and thus antigen concentration while holding the antibody level constant (Fig. 3-12) or by the successive addition, incubation, and centrifugation of fresh cells to subsequent supernatants and summing the total antibody bound (relative to a nonspecific control IgG). It is also quite desirable to examine the performance of an internally labeled antibody under the identical assay conditions used for iodinated or ^{111}In-labeled antibody. This can be achieved by incubation of hybridoma cultures with [^{35}S]methionine or [^{14}C]leucine (Lampson, 1980) and purification of the labeled monoclonal antibody from the cell culture supernatants. It is then possible to determine the maximum binding for a given antibody under optimal conditions and what the effect of iodination or chemical modification is on the magnitude of binding. This is of obvious importance to the use of labeled monoclonals for imaging studies, since labeled but nonreactive antibody could contribute significantly to the body background.

Unfortunately, few of the published studies on the use of radiolabeled

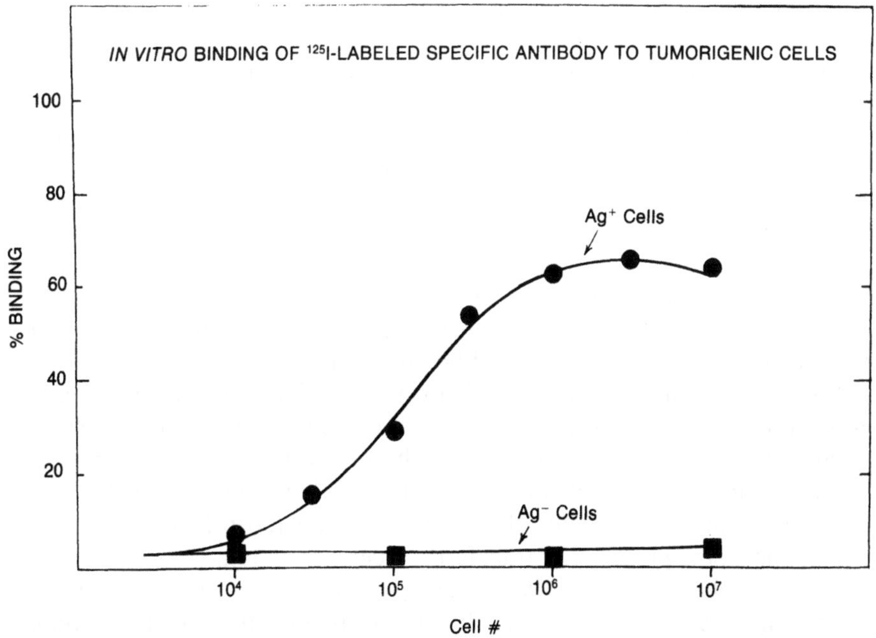

FIG. 3-12. Generalized quantitative binding assay for the determination of immunoreactivity.

antibodies, whether polyclonal or monoclonal, include adequate or appropriate controls. The use of isotype-matched nonspecific monoclonals is a reasonable control to differentiate localization due to Fc binding to cells expressing receptors to this domain on immunoglobulins. However, simple changes in capillary permeability and the degree of vascularity of tumors (Folkman, 1976) should also be explored in all imaging and localization studies. This has been investigated by investigators far too infrequently and deserves some mention here. The use of labeled human serum albumin (HSA) or similar vascular indicator has been largely confined to the application of blood pool subtraction techniques to differentiate activity in the tumor due to blood-borne activity from extravascular localization. Considerable controversy exists still on the relative merits and drawbacks of the use of computer subtraction techniques to enhance the tumor to background ratios. The extravascular and extracellular space of most tumors is significantly greater than that for normal tissues (Bale et al., 1980). This results in the slow but significant diffusion of proteins such as antibodies into this extracellular space. To determine the specific antibody accumulation, the use of a second, nonspecific antibody or labeled albumin labeled with a different radionuclide is injected into the same individual. The computer is used to subtract one image from the other, or in this example, the nonspecific antibody distribution from the specific. However, the frequently low tumor-to-background ratios observed can easily result in positive artifacts using this technique (Mach et al., 1980). Therefore, it is important when computer subtraction techniques are to be employed that the data presented include the unprocessed specific antibody image, the unprocessed nonspecific antibody or control protein image, and finally the subtracted tumor image. Unfortunately, the unprocessed blood pool image is almost never shown; rather, the antibody picture and the blood pool subtracted pictures are presented. It is of great significance and relevance to know if the tumor could also be visualized by the blood pool image taken with HSA alone. If the degree of vascularity and/or permeability within the tumor are sufficient for detection using currently available radiopharmaceuticals, such as labeled HSA, then the application of a labeled antibody, whether monoclonal or polyclonal, for this purpose should be questioned.

Some studies have used either ^{111}In or ^{113m}In-labeled transferrin as the vascular indicator. It has recently been shown that ^{113m}In-labeled transferrin is not a satisfactory substitute for [^{125}I]transferrin or [^{125}I]albumin in the measurement of transcapillary permeability (Nahmias et al., 1981). This is due to a significantly greater clearance of In-transferrin from the plasma. Furthermore, the presence of transferrin receptors on tumors (Larson et al., 1980; Shindelman et al., 1981; Goding and Burns, 1981; Omary and Trowbridge, 1981; Wilson et al., 1982) may potentially complicate tumor localization due to immunorecognition phenomena from alternative localization mechanisms.

Perhaps the single most omitted control, however, is the use of [^{67}Ga]-citrate. [^{67}Ga]citrate is accumulated by a number of animal and human tumors following intravenous administration (Edwards and Hayes, 1969; Edwards and Hayes, 1970; Larson, 1978; Noujaim et al., 1979; Wong et al., 1980; Hoffer, 1980). Although this radiopharmaceutical is not ideal in terms both of its physical and pharmacological properties for tumor imaging, it is currently the agent of choice for tumor imaging. It is essential, therefore, that experiments carried out in experimental tumor models that proclaim the potential clinical utility of labeled monoclonals include suitable comparisons with the current clinical agent of choice. This will add relevance to the results presented and allow a more realistic evaluation of the true clinical potential of a given monoclonal antibody. A recent study (Colcher et al., 1982) that has incorporated many of the appropriate controls into the experimental protocol has investigated the use of a monoclonal antibody, B6.2 reactive with a protein found on the surface of both primary and metastatic human mammary tumors (Colcher et al., 1981). The intact IgG as well as $F(ab')_2$ and Fab′ fragments were radiolabeled with ^{125}I and shown to retain their immunoreactivity and specificity (Fig. 3-13). All were successful in localizing human mammary tumors transplanted into athymic mice with tumor-to-tissue ratios improving over a 4-day period. The $F(ab')_2$ fragment gave higher tumor-to-tissue ratios than the intact IgG. No localization of labeled B6.2 or its fragments were observed in athymic mice bearing a human melanoma (not expressing the target antigen), nor was there any localization of isotype-matched control immunoglobulins or their fragments into mammary tumor. Imaging studies confirmed the ability of radiolabeled B6.2 to detect tumors of 0.4-cm diameter without the use of computer subtraction methods (Figs. 3-14 and 3-15).

The Nude Mouse

In 1962 a hairless mouse mutant was found in the Virus Laboratories, Ruchill Hospital in Glasgow, Scotland, and subsequently bred until first described in detail by Flanagan in 1966 (Flanagan, 1966). This was soon followed by the discovery that these animals lacked the thymus gland (Pantelouris, 1968). This condition results in the inability of the nude mouse to recognize biological nonself in that cell-mediated immune functions are severely impaired. The standard measure of cell-mediated immunity at that time was transplantation reactions, especially reactions to skin grafts (Rygaard, 1978). In the summer of 1969 a sufficient number of nude mice were investigated for rejection of rat skin grafts that were readily and permanently accepted (Rygaard, 1969). These and subsequent findings confirmed the expectation that nude mice were immunodeficient and could readily accept heterotransplants. These early findings were soon extended to a human malignant tumor of the colon that could be established and serially trans-

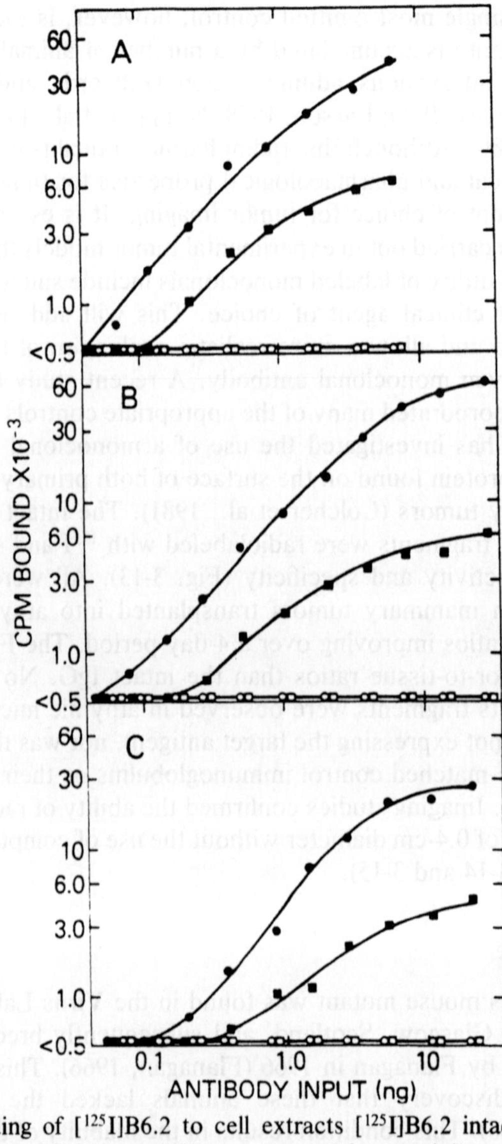

FIG. 3-13. Binding of [^{125}I]B6.2 to cell extracts [^{125}I]B6.2 intact IgG (panel A), F(ab')$_2$ (panel B) and Fab' (panel C) were assayed for reactivity to cell extracts from a human breast tumor metastasis to the liver (closed circles), the MCF-7 breast tumor cell line (closed squares), normal human liver (open circles), and the A204 rhabdomyosarcoma cell line and the NC37 human lymphoid cell line (open squares). The antibody added was adjusted for the different specific activities to the three preparations. (From Colcher et al., 1983.)

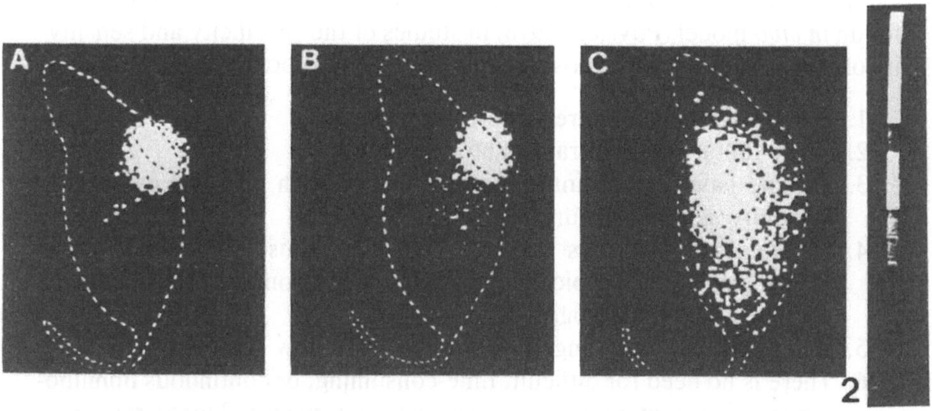

FIG. 3-14. Gamma camera images of [^{125}I]B6.2 in athymic mice bearing a 0.5-cm transplantable human mammary tumor (Clauser, panels A and B) or a human melanoma (A375, panel C). The mice were scanned at 24 hours (panels A and C) and 96 hours (panel B) until an equal number of counts was accumulated. (From Colcher et al., 1983.)

planted in these animals (Rygaard et al., 1969). The histological appearance of the tumor remained similar to the original implants after about 70 passages over an 8-year period. These types of findings have been subsequently confirmed in many other laboratories (Giovanella et al., 1972; Sordat et al., 1974; Shimosato et al., 1974).

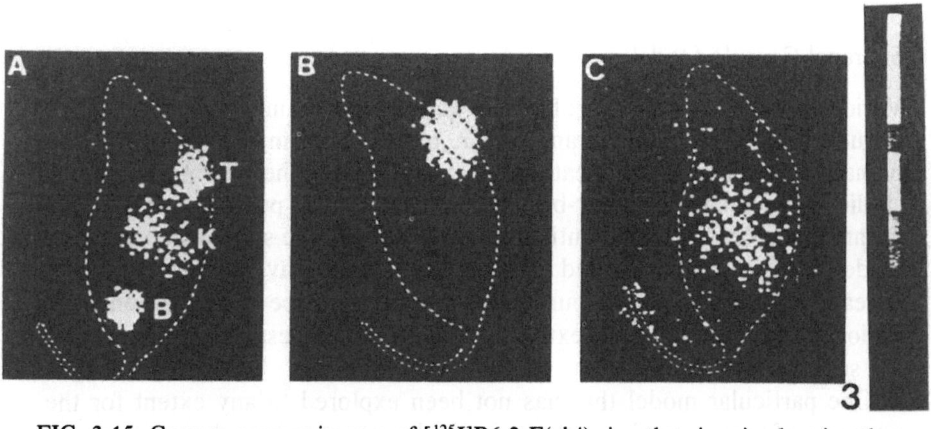

FIG. 3-15. Gamma camera images of [^{125}I]B6.2 F(ab')$_2$ in athymic mice bearing the Clauser human mammary tumor (panels A and B) or A375 human melanoma (panel C) at 24 hours (panel A) or 96 hours (panels B and C) for an equal number of counts. (From Colcher et al., 1983.)

The human tumor-nude mouse system fulfills a number of requirements for an *in vivo* model (Povlsen, 1978) in studies of the specificity and sensitivity of human tumors to diagnosis and/or therapy with monoclonal antibodies:

1. Human solid tumors are readily accepted.
2. Many tumors can be transplanted serially.
3. Tumors have a constant and predictable growth rate that can be followed by simple measurements.
4. Human characteristics are preserved in mouse-grown tumors as judged from microscopic examination, chromosome analysis, isozyme studies, and immunological investigations.
5. Extensive and long-range investigations are now possible.
6. There is no need for difficult, time-consuming, or continuous immunosuppressive treatment (e.g., cortisone) that might influence results.

A number of factors will influence the transplantation of various types of tumor tissue into nude mice (Reid et al., 1978). These include

a. The health of the animals.
b. The handling and preparation of surgical specimens.
c. The site of implantation.
d. Appropriate hormonal conditioning of the host.
e. The time factor in determining how long to wait for the development of a tumor line.
f. Whether the tumor is benign or malignant.
g. The ability of the tumor to be vascularized by the host.
h. The ability of nude mice to recognize particular foreign antigens and to suppress the tumorigenicity of cells carrying the antigens.

Subrenal Capsule Model

Various sites of implantation of human xenografts into mice have been tried, including subcutaneous, intramuscular, intrasplenic, in the footpad, intrapinnal, intracranial, and beneath the kidney capsule. The vast majority of all studies carried out in tumor-bearing animals for the purposes of external scanning with radiolabeled antibodies have utilized the subcutaneous tumor model. However, as indicated earlier, these studies have usually employed unrealistically large tumor burdens; in addition, since target-to-nontarget radioactivity is essential for external detection, the question of relevancy of the subcutaneous site arises.

One particular model that has not been explored to any extent for the purposes of tumor imaging is the recently developed subrenal capsule model (Bogden et al., 1978). The objective in developing this model was for rapid drug screening of potential new chemotherapeutic agents using human tumor xenografts for several reasons:

a. To circumvent the variable of immunocompetency of the xenograft host.
b. To use fewer animals and less time (at reduced cost).
c. To provide a realistic screen against a clinically relevant tumor burden in an effort to mimic more closely the postsurgery residual or metastatic tumor for evaluating drug activity.

Briefly described, preselected 1-mm^3 fragments of viable tumor tissue are implanted under the renal capsule, which provides a rich vascular bed for nutrient as well as drug delivery. The size of the implant is measured with a stereoscopic microscope fitted with an ocular micrometer so that 10 ocular micrometer units equal 1 mm in diameter (Fig. 3-16). Measurement of the initial and final tissue graft sizes *in situ,* in control versus drug-treated animals over an 11-day assay period, provides the parameter for the ranking of various chemotherapeutic agents. The growth rates of five human tumors (two colon, two mammary, and one lung) tested in this subrenal capsule assay were significant enough to permit quantiation of chemotherapeutic agents within the 11-day period (Fig. 3-17) (Bogden et al., 1978).

When subcutaneous and subrenal capsule assay procedures were compared by the evaluation of test results of 112 compounds using all three primary xenograft panel tumors of the National Cancer Institute (MX-1 breast, CX-1 colon, and LX-1 lung) and applying an activity calculation method common to both assay procedures, it was found that the subrenal capsule assay was more likely to detect potential activity than the subcutaneous assay (Bogden et al., 1982). More recently, the subrenal capsule assay has been shortened to a 6-day assay using normal immunocompetent mice (Bogden et al., 1979; Bogden et al., 1981). This was possible because

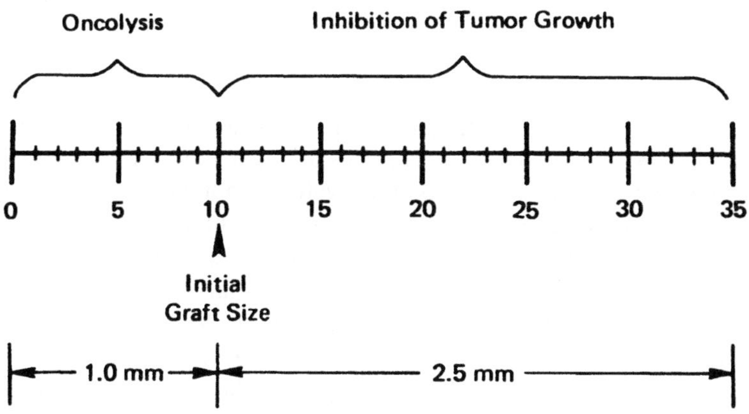

FIG. 3-16. Scale of ocular micrometer units (OMU) compared to millimeters used for measuring chemotherapeutic effects in the subrenal capsule assay. (From Bogden et al., 1978.)

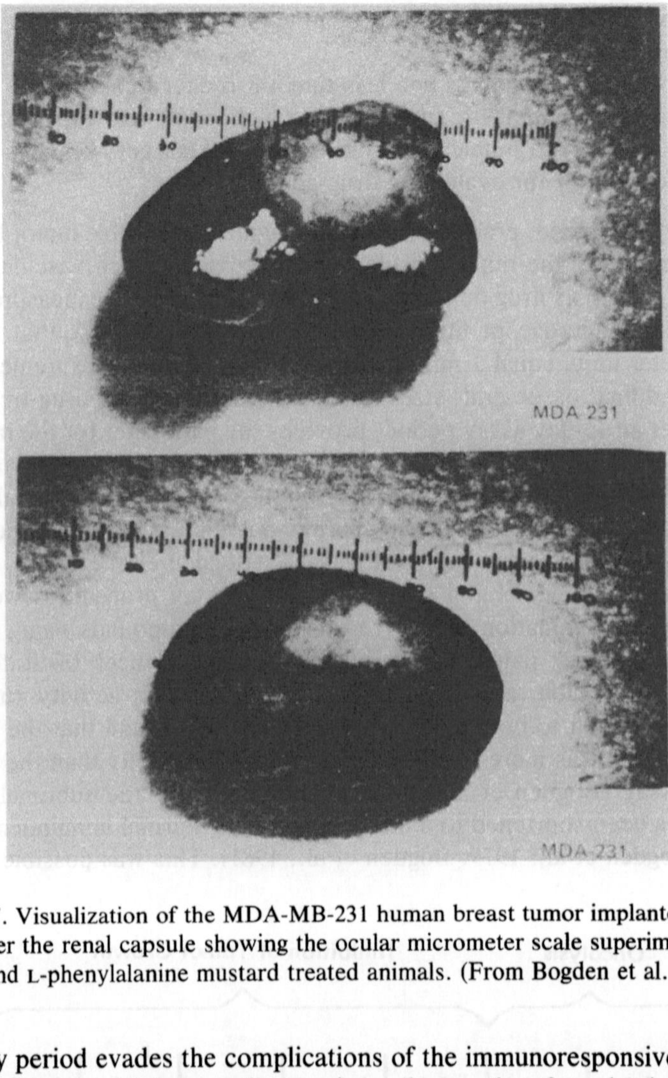

FIG. 3-17. Visualization of the MDA-MB-231 human breast tumor implanted for 11 days under the renal capsule showing the ocular micrometer scale superimposed in control and L-phenylalanine mustard treated animals. (From Bogden et al., 1978.)

the 6-day period evades the complications of the immunoresponsiveness of the normal host, which normally requires a longer time for rejection.

The extension of this model to investigate the potential of using radiolabeled monoclonal antibodies has been recently demonstrated using a solid murine lymphoma growing under the subrenal capsule of normal, immunocompetent mice (Sands et al., 1983). This model is far more representative of human metastases as tumors of approximately 20–60 mg could be specifically detected. The experimental system comprised the use of two lymphoma lines, one expressing Thy 1.1 antigen on their surface (SL2) and the other being Thy 1.1 negative (SL1). A monoclonal antibody, MRC OX7 specifically directed against the Thy 1.1 antigen (Mason et al., 1980) and an isotype matched (IgG_1) nonspecific immunoglobulin, antihorseradish peroxidase (HRP), were radioiodinated and injected into animals containing either

SL2 or SL1 tumors. It should be noted that the SL2 tumor (Thy 1.1^+) growing in the AKR Cumberland strain of mouse, which lacks the Thy 1.1 antigen, represents a model of a "tumor-specific" antigen.

Eight days after implantation of 1 mm^3 explants of either tumor type under the renal capsule, animals were injected with both antibodies to determine the specificity of localization *in vivo* (*in vitro* binding studies had shown less than 1% binding of radiolabeled MRC OX7 to SL1 cells and greater than 70% binding in a single incubation with SL2 cells). The specific uptake of OX7 into the SL2 tumors was clearly evidenced and showed about a six-fold greater absolute uptake (% injected dose/g) (Fig. 3-18). Neither SL1 nor SL2 tumors showed accumulation of the anti-HRP. Figure 3-19 shows the results expressed in terms of localization index (LI). The LI for SL2 tumor in the subrenal capsule reached 12 by 3 hours and over 15 at 48 hours. The LI for OX7 in the SL 1 tumor remained approximately 1, indicating no specific antibody accumulation in antigen negative (control) tumors.

Imaging studies in animals bearing SL2 tumor in their right kidney and SL1 tumor in their left kidney and injected with either ^{131}I OX7 or anti-HRP show the selective accumulation of OX7 in the kidney bearing the small SL2 tumor (Fig. 3-20). The OX7 did not visualize the nonspecific SL1 tumor and the anti-HRP did not image either tumor. In addition, ^{67}Ga-citrate failed to image either tumor (Sands et al., 1983). These data support the use of more clinically relevant evaluation systems under well-controlled experimental

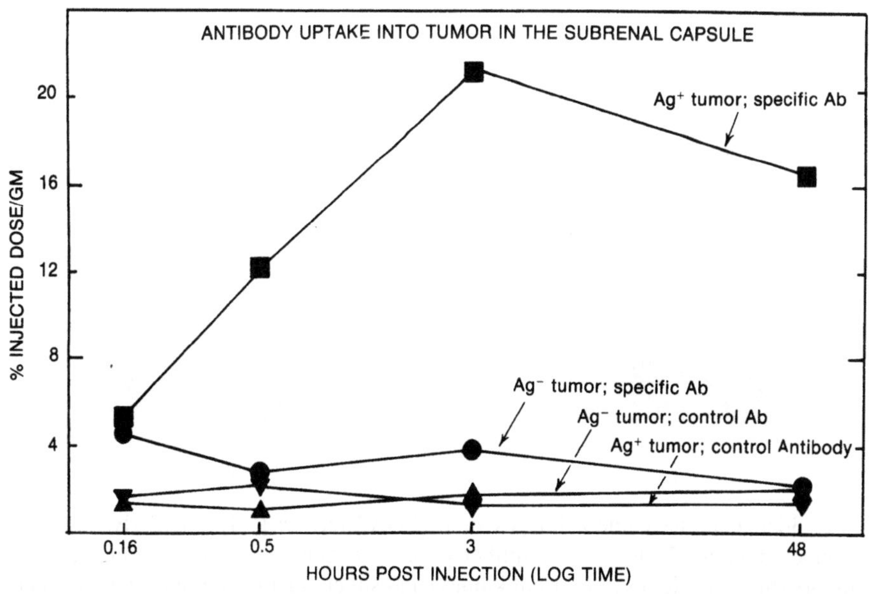

FIG. 3-18. Biokinetics of [^{125}I]MRC-OX7 (specific Ab) or [^{131}I]antihorseradish peroxidase (control Ab) in AKR/Cumberland mice bearing SL2 (antigen positive) versus SL1 (antigen negative) tumors of 20–60 mg under the renal capsule over 48 hours.

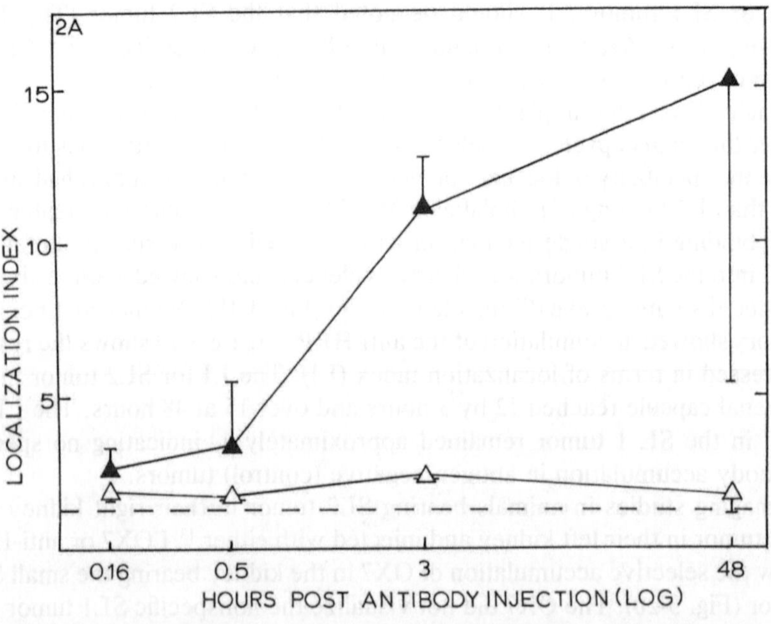

FIG. 3-19. Localization index over 48 hours [^{125}I]MRC-OX7 into SL2 (antigen positive) tumors ▲-▲ compared to [^{131}I]anti HRP △-△ (a control isotype matched antibody) growing under the renal capsule.

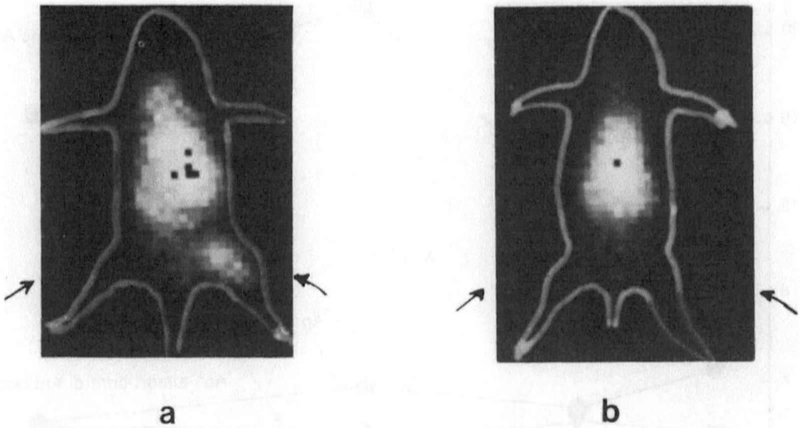

FIG. 3-20. Gamma camera images of [^{125}I]OX7 (A) or [^{131}I]-αHRP (B) in AKR/Cumberland mice 24 hours after antibody injection. Both mice have subrenal capsule tumors (SL1 left kidney; SL2 right kidney) and subcutaneous tumors (SL1 left thigh; SL2 right thigh). SL2 tumor in the right kidney shows highest intensity pixels in A, whereas no assymetry was observed in B. Arrows show lack of [α-^{131}I]HRP uptake in either subcutaneous tumor type and low specific uptake of [^{125}I]OX-7 in subcutaneous SL2 tumor.

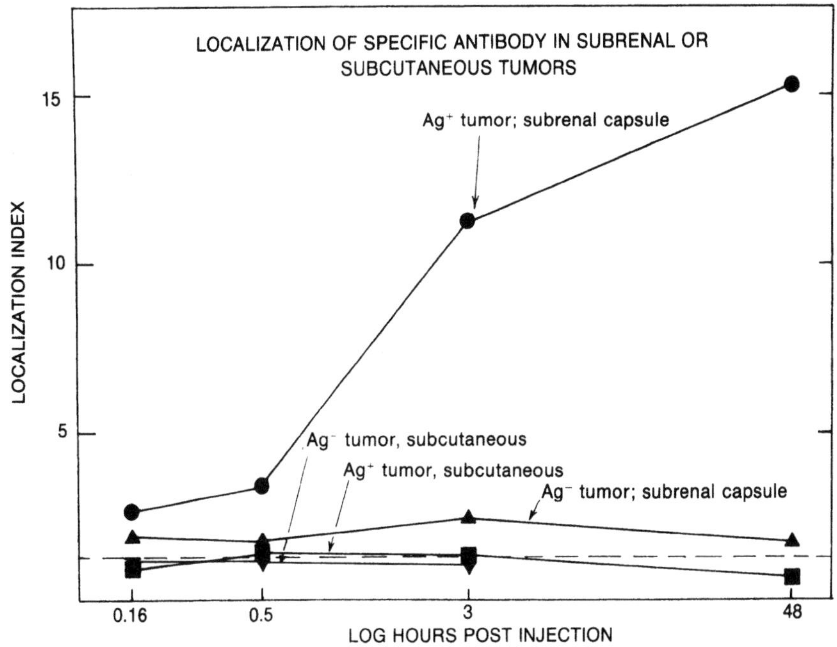

FIG. 3-21. Localization index as a function of time in SL2 (Ag^+ tumor) versus SL1 (Ag^- tumor) at two different implantation sites.

conditions. This is the first report of the experimental detection of tumors of this size (20–60 mg).

Additional studies carried out in the same model system have compared subrenal capsule tumors to subcutaneously grown tumors (Sands et al., 1983). Quite surprisingly, the subcutaneous tumors showed greatly decreased specificity of antibody accumulation compared to the subrenal capsule tumors (Fig. 3-21). At present, the explanation of these results is unclear, although it is speculated that a greater macrophage infiltration of tumors grown of the subcutaneous site may be involved. The results may also reflect differences in tumor blood flow. These recent findings reinforce the potential importance of the model system chosen for the evaluation of radiolabeled monoclonal antibodies. From an intuitive standpoint, it is tempting to speculate that model systems such as the subrenal capsule model may be more predictive of what one might expect in a clinical setting, but this relationship has yet to be demonstrated.

Metastatic Models

Other metastatic models of human disease clearly need to be investigated at present. Only relatively infrequent reports of metastases of human tumors in nude mice had been reported until 1979 (Giovanella et al., 1974; Giovanella et al., 1973; Shimosato et al., 1976; Sharkey and Fogh, 1979). More recent

TABLE 3-9. Growth of HE_p-2 Cells in Nude Mice[a]

No. Animals Injection Site	*No. Cells Injected*	*Mitotic Activity*	*Supporting Stroma*	*Necrosis*	*Invasiveness*	*Metastasis*	*Tumor wt (g)*
5/SC	10^6	High	++[b]	++++	+	None	0.419 (0.194–0.919)
5/SC	2×10^7	High	++	++++	+	None	0.874 (0.194–0.918)
7/IP	10^6	High	±	++	++++	3/7	2.152 (1.643–2.818)
7/IP	2×10^7	High	±	++	++++	3/7	2.548 (1.906–3.407)

[a] From Kyriazis et al. (1978).
[b] Intensity of findings from mild (+) to marked (++++).

studies, however, have demonstrated that various human tumors transplanted into athymic mice are characterized by progressive growth, local invasion, and distant metastases primarily to lungs and mediastinal structures and that the site of implantation is important in the expression of these characteristics (Kyriazis et al., 1978; Kyriazis et al., 1978; Kyriazis et al., 1980). This has been clearly demonstrated for two human tumor cell lines, HEp-2 derived from a carcinoma of the larynx and SW480 derived from a colon carcinoma (Kyriazis et al., 1978). Athymic mice were injected either subcutaneously or intraperitoneally with either 10^6 of 2×10^7 cells. The results of these studies are shown in Tables 3-9 and 3-10. Both routes of

TABLE 3-10. Growth of SW480 Cells in Nude Mice[a]

No. Animals Injection Site	*No. Cells Injected*	*Mitotic Activity*	*Supporting Stroma*	*Necrosis*	*Invasiveness*	*Metastasis*	*Tumor wt (g)*
5/SC	10^6	High	++[b]	++++	+	None	0.942 (0.304–2.881)
5/SC	2×10^7	High	++	++++	+	None	0.940 (0.533–1.655)
6/IP	10^6	High	±	+±	++++	3/6	ND
6/IP	2×10^7	High	++	++	++++	6/6	ND

[a] From Kyriazis et al. (1978).
[b] Intensity of findings from mild (+) to marked (++++).

transplantation resulted in tumor growth. However, growth rates, invasiveness, and metastatic behavior of the transplanted tumors differed, depending on the route of transplantation. A characteristic feature of the subcutaneous tumors was their complete encapsulation with fibrous tissue, which may inhibit growth and/or local invasiveness and metastases. It may also affect the vascularization of the implant and contribute to the much greater necrosis observed subcutaneously. In contrast, the intraperitoneal tumors grew in a multifocal pattern throughout the peritoneal cavity and were characterized by minimal necrosis. Tumor masses were mostly not encapsulated and extensive metastases to the mediastinal lymph nodes and lungs were observed. These findings suggest that intraperitoneally grown human tumors may more closely imitate the histological characteristics of the malignant growth in man expressed by their tendency to spread over a wide area, infiltrate, and give distant metastases by both lymphatic and hematogeneous routes (Kyriazis et al., 1978).

Published reports have indicated that tumors originating in various strains of conventional mice showed marked regional growth differences when transplanted intradermally into their syngeneic host (Auerbach et al., 1978a, 1978b). A similar pattern has been observed for subcutaneously grown human tumors in nude mice (Kyriazis and Kyriazis, 1980). It was observed that a variety of human tumors transplanted in the anterior lateral thoracic wall (Fig. 3-22) grew faster than did tumors transplanted in the posterior aspect of the trunk. Anteriorly growing tumors were better vascularized and less necrotic and showed greater invasiveness and greater incidences of metastases to the regional lymph nodes (Table 3-11). These findings were independent of tumor type studied. It was concluded that regional vascular supply may be a

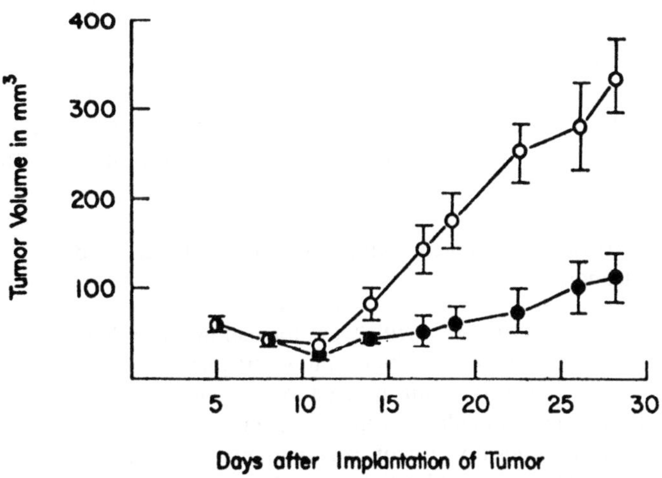

FIG. 3-22. Growth curves of a human pancreatic adenocarcinoma as a function of implantation site. ○, anteriorly grown; ●, tumors growing posteriorly. (From Kyriazis and Kyriazis, 1980.)

TABLE 3-11. Comparison of Regional Biological and Histopathological Differences of Human Tumors Grown Subcutaneously in Nude Mice[a,b]

Site of Growth	*Growth Rate*	*Vascularity*	*Necrosis*	*Fibrosis*	*Mitotic Activity*	*Local Tissue Infiltration*	*Metastases*
Anterior Trunk region	++++[c]	+++	+	±	++	++	Observed
Posterior Trunk region	++	++	++	+	++	– –	– –

[a] From Kyriazis and Kyriazis (1980).
[b] Tumor characteristics 30 days posttransplantation.
[c] Intensity of findings from mild (+) to marked (+ + + +); not seen (– –).

key factor influencing the biological behavior of such transplanted tumors (Kyriazis and Kyriazis, 1980). These findings should alert those pursuing radiolabeled antibodies for tumor detection as to the possible effects such differences in tumor biology may have on localization. Again this parameter has been virtually ignored by those pursuing *in vivo* tumor detection. It is of utmost importance in the design of animal experimental models that the real potential clinical needs and problems are addressed. Table 3-12 lists some of the current problems perceived in the use of monoclonal antibodies for tumor diagnosis and therapy. Many of these points have been discussed earlier; however, a recent paper has illustrated the effect of circulating host antibodies in a murine system to emphasize the potential significance of this problem in experimental models (Buchsbaum et al., 1982). In this study groups of animals were implanted with an SV40-transformed mouse tumor expressing SV40 tumor virus associated antigens. Groups were either immunosuppressed by prior X-irradiation and cortisone treatment (to prevent

TABLE 3-12. Problems with MAb Tumor Therapy and Diagnosis

1. Specificity (how specific? less for therapy than diagnosis)
2. Pharmacokinetics (which isotope, fragments, conjugates)
3. Circulating free antigen (need to saturate)
4. Antigenic modulation (dose regimens)
5. Antigen negative tumor cells (mutants, metastases)
6. Circulating host blocking antibodies
7. Immune response of host (toxicity)
8. Tumor sanctuary sites

the host animal from mounting an immune response to the foreign antigens on the tumor surface) or nonimmunosuppressed mice. Polyclonal anti-SVT2 cells mouse immunoglobulins were radioiodinated and injected into the tumor-bearing control or immunosuppressed animals. The labeled antibodies showed a significantly greater preferential localization in the immunosuppressed compared with control animals. Furthermore, the localization of antibody into the SVT2 tumor in immunosuppressed mice could be reduced significantly by the passive administration of anti-SVT2 serum (Buchsbaum et al., 1982). These findings merely illustrate another factor that can significantly influence localization of radiolabeled antibodies.

SUMMARY

The availability of highly purified and specific monoclonal antibodies offers tremendous potential. This is true for many of the possible applications discussed earlier, not the least of which is for the radioimmunodetection of cancer. It is essential, however, that systematic and well-controlled studies of this potential be carried out in clinically relevant animal model systems. Many of the potential pitfalls as well as recommendations for the future have been discussed. In summary, the recommendations for the evaluation of monoclonal antibodies *in vitro* and *in vivo* are listed in Table 3-13.

The use of nonspecific controls to rule out localization due to nonspecific (Fc) binding or vascular permeability changes and the direct comparison to the current clinical agent of choice, [^{67}Ga]-citrate, should be a part of every monoclonal antibody investigation. If computer subtraction methods are to be used, show the original antibody image, the original blood pool image, and the subtracted image. It is essential that the optimal immunoreactivity be determined using endogenously labeled (e.g., [^{14}C], [^{3}H], or [^{35}S]) antibody under antigen excess conditions so that the effects of radioiodination or labeling with bifunctional chelators can be determined prior to animal experimentation. It is of significance to demonstrate that antibody binding to tumor cell membranes is indeed reflective of immunorecognition due to binding of the antigen combining site and not nonspecific adsorption. This can be done in a number of ways in comparison with isotype-matched controls and tumor cells not expressing the specific antigens in question.

A frequently ignored parameter is the demonstration that the antigen against which a particular monoclonal antibody is directed *in vitro* continues to be expressed on the target cells after growing for several days to weeks *in vivo*. Additionally, consideration for how this expression may change as a function of tumor age, size, and expression in metastases should be made. The localization of radiolabeled antibodies in true metastatic models is an almost totally unexplored area for animal investigation and represents perhaps the most clinically relevant evaluation system. With some of the newer information concerning metastases of human xenografts in nude animals, it

TABLE 3-13. Recommendations for the Evaluation of MAb *In Vitro* and *In Vivo*

1. Nonspecific controls
   a. HSA
   b. IgG (nonspecific, isotype matched)
   c. [^{67}Ga]citrate
   d. Polyclonal antiserum
   e. [^{111}In]transferrin (may not be negative control)
2. For computer substraction studies show
   a. Original antibody image
   b. Original blood pool image
   c. Subtracted tumor image
3. Quantitative assays for immunoreactivity
   a. ^{14}C/^{35}S/^{3}H endogenously labeled-% maximum binding under antigen excess conditioning
   b. *I-labeled—% maximum binding
   c. [^{111}In]DTPA, EDTA, etc.—% maximum binding
4. Demonstrate antibody binding to tumor cell membranes
   a. Fluorescence techniques (FACS, second antibody techniques, etc.)
   b. Autoradiography
5. Demonstration of antigen expression *in vivo*
   a. As a function of tumor age
   b. As a function of tumor size
6. Demonstration of localization in metastases
7. Variations in implantation sites
   a. Subcutaneous
   b. Intramuscular
   c. Subrenal capsule
   d. Liver, lung, pleural cavity
   e. Brain
8. Check for circulating antigen and the production of host antibody as tumor grows and necrosis begins
9. Check nonspecific tumors in the same site
10. Investigate the relationship between uptake specificity and tumor growth rate and tumor size.

should be possible to develop such metastatic models for future investigations.

Some of the potential pitfalls of implantation sites have been illustrated. It is essential, therefore, that alternative implantation sites for a given tumor be investigated to mimic more closely the likely sites of tumor in the human and because the biology of a given tumor type can vary markedly as a function of growth environment. The possible influences of shed antigen into the extracellular fluid and blood as well as the potential for the production of host autoantibodies should be considered and measured in well-controlled studies of antibody localization. The nonspecific accumulation due to an increased permeability of tumor endothelium should be measured both by using nonspecific controls, such as nonimmune IgG of the same isotype, and by using nonspecific tumors (i.e., not expressing the specific antigen under investigation) with similar morphological and biological characteristics.

Finally, the relationship between tumor uptake of monoclonal antibody and tumor growth rate and size may be critical to the utility of this approach for earlier clinical diagnosis. This parameter should be investigated wherever possible to provide a more realistic appraisal of the concept of using radiolabeled antibodies for this purpose. It should be clear that the future of these materials is very bright indeed but that there remains a significant challenge to the researcher to design and evaluate this potential in a truly rigorous and controlled fashion.

REFERENCES

Anzel, L. M., and Poljak, R. J. (1979). Three-dimensional structure of immunoglobulins. *Ann. Rev. Biochem.* **48:**961–983.

Ashall, F., Bramwell, M. E., and Harris, H. (1982). A new marker for human cancer cells. 1. The Ca antigen and the Ca1 antibody. *Lancet* **3**(July):1–6.

Atassi, M. Z. (1979). *Immunochemistry of Proteins.* New York: Plenum Press.

Auerbach, R., Morrissey, L. W., and Sidky, Y. A. (1978a). Gradient in tumor growth. *Nature* **274:**697–699.

Auerbach, R., Morrissey, L. W., and Sidky, Y. A. (1978b). Regional differences in the incidence and growth of mouse tumors following intradermal or subcutaneous inoculation. *Cancer Res.* **28:**1739–1744.

Bagshawe, K. D., Searle, F., Lewis, J., Brawn, P., and Keep, P. (1980). Preliminary therapeutic and localization studies with human chorionic gonadotropin. *Cancer Res.* **40:**3016–3017.

Bale, W. F., Contreras, A. M., and Grady, E. D. (1980). Factor influencing localization of labeled antibodies in tumors. *Cancer Res.* **40:**2965–2972.

Ballou, B., Levine, G., Hakala, T. R., and Solter, D. M. (1979). Tumor location detected with radioactivity labeled monoclonal antibody and external scintigraphy. *Science* **206:**844–847.

Bodgen, A. F., Kelton, D. E., Cobb, W. R., and Esber, H. J. (1978). A rapid screening method for testing chemotherapeutic agents against human tumor xenografts. In *Proceedings of the Symposium on the Use of Athymic (Nude) Mice in Cancer Research,* Houchens, D. P., and Ovejera, A. A., eds. Stuttgart: Gustav Fischer.

Bodgen, A. E., Haskell, P. M., LePage, D. J., Kelton, D. E., Cobb, W. R., and Esber, H. J. (1979). Growth of human tumor xenografts implanted under the renal capsule of normal immunocompetent mice. *Exp. Cell Biol.* **47:**281–293.

Bodgen, A. E., Cobb, W. R., LePage, D. J., Haskell, P. M., Gulkin, T. A., Ward, A., Kelton, D. E., and Esber, H. J. (1981). Chemotherapy responsiveness of human tumors as first transplant generation xenografts in the normal mouse: Six day subrenal capsule assay. *Cancer* **48:**10–20.

Bogden, A. E., Houchens, D. P., Ovejera, A. A., and Cobb, W. R. (1982). Advances in chemotherapy studies with nude mice. In *The Nude Mouse in Experimental and Clinical Research,* Vol. 2, Fogh, J., and Giovanella, B. C., eds. New York: Academic Press.

Buchsbaum, D. J., Anderson, J. M., and Bray, B. E. (1982). Localization of radiolabeled antibody in SVT2 tumor increases with immunosuppression of the host. *Proc. Soc. Exp. Biol. Med.* **171:**56–64.

Burchiel, S. W., Khaw, B. A., Rhodes, B. A., Smith, T. W., and Haber, E. (1982). Immunopharmacokinetics of radiolabeled antibodies and their fragments. In *Tumor Imaging: The Radioimmunochemical Detection of Cancer,* Burchiel, S. W. and Rhodes, B. A. New York: Masson Publishing.

Burnet, M. (1959). *The Clonal Selection Theory of Acquired Immunity.* Nashville: Vanderbilt University Press.

Butler, V. P., Watson, J. F., Schmidt, D. H., Gardner, J. D., Mandal, W. J., and Skelton, C. L. (1973). Reversal of the pharmacological and toxic effects of cardiac glycoside by specific antibodies. *Pharmacol. Rev.* **25:**239–248.

Carey, T. E., Takahashi, T., Resnick, L. A., Oettgen, H. F., and Old, L. J. (1976). Cell surface antigens of human malignant melanoma. 1. Mixed hemadsorption assays for humoral immunity to cultured autologous melanoma cells. *Proc. Natl. Acad. Sci. USA* **73:**3278–3282.

Colcher, D., Hand, P. H., Nuti, M., and Scholm, J. (1981). A spectrum of monoclonal antibodies reactive with human mammary tumor cells. *Proc. Natl. Acad. Sci. USA* **78:**3199–3203.

Colcher, D., Zalutsky, M., Kaplan, W., Kufe, D., Austin, F., and Scholm, J. (1983). Radiolocalization of human mammary tumors in athymic mice by a monoclonal antibody. *Cancer Res.* **43:**736–742.

Crawford, L.V., Pim, D. C., Gurney, E. G., Goodfellow, P., and Taylor-Papadimitriou, J. (1981). Detection of a common feature in several human tumor cell lines a 53,000-dalton protein. *Proc. Natl. Acad. Sci. USA* **78:**41–45.

Cuttitta, F., Rosen, S., Gazdar, A. F., and Minna, J. D. (1981). Monoclonal antibodies that demonstrate specificity for several types of human lung cancer. *Proc. Natl. Acad. Sci. USA* **78:**4591–4595.

Day, E. D., Planinsek, J. A., and Pressman, D. (1959). Localization *in-vivo* of radioiodinated antifibrin antibodies to Murphy rat lymphosarcoma and other tumors. *J. Natl. Cancer Inst.* **22:**413–426.

Eckelman, W. C., Paik, C. H., and Reba, R. C. (1980). Radiolabeling of antibodies. *Cancer Res.* **40:**3036–3042.

Eckelman, W. C., Reba, R. C., Gibson, R. E., Rzeszotarski, W. J., Vieras, F., Mazaitis, J. K., and Francis, B. (1979). Receptor binding radiotracers: a class of potential radiopharmaceuticals. *J. Nucl. Med.* **20:**350–357.

Edwards, C. L., and Hayes, R. L. (1969). Tumor scanning with ^{67}Ga-citrate. *J. Nucl. Med.* **10:**103–105.

Edwards, C. L., and Hayes, R. L. (1970). Scanning malignant neoplasms with ^{67}Gallium. *JAMA* **212:**1182–1184.

Eisen, H. N. (1980). *Immunology.* New York: Harper & Row.

Embleton, M. J., Gunn, B., Byers, V. S., and Baldwin, R. W. (1981). Antitumor reactions of monoclonal antibody against a human osteogenic sarcoma cell line. *Brit. J. Cancer* **43:**582–587.

Fidler, I. J. (1978). Tumor heterogeneity and the biology of cancer invasion and metastasis. *Cancer Res.* **38:**2651–2658.

Fidler, I. J., and Hart, I. R. (1981). The origin of metastatic heterogeneity in tumors. *Eur. J. Cancer* **17**:487–494.

Flanagan, S. P. (1966). "Nude," a new hairless gene with pleiotropic effects in the mouse. *Genet. Res.* **8**:295–309.

Fogel, M., Gorelik, E., Segal, S., and Feldman, M. (1979). Differences in cell surface antigens of tumor metastases and those of local tumor. *J. Natl. Cancer Inst.* **62**:585–596.

Folkman, J. (1976). The vascularization of tumors. *Sci. Am.* **234**:58–73.

Frankel, M. E., and Gerhard, W. (1979). The rapid determination of binding constants for antiviral antibodies by a radioimmunoassay. An analysis of the interaction between hybridoma proteins and influenza virus. *Mol. Immunol.* **16**:101–106.

Geran, R. I., Greenberg, N. H., MacDonald, M. M., Schumacher, A. M., and Abbott, B. J. (1972). Protocols for screening chemical agents and natural products against animal tumors and other biological systems, 3rd ed. *Cancer Chemother. Rep.* **3**(2):51.

Giovanella, B. C., Yim, S. O., Stehlin, J. S., and Williams, L. J. (1972). Development of invasive tumors in the nude mouse after injection of cultured human melanoma cells. *J. Natl. Cancer Inst.* **48**:1531–1533.

Giovanella, B. C., Yim, S. O., Morgan, A. C., Stehlin, J. S., and Williams, L. J. (1973). Metastases of human melanomas transplanted in nude mice. *J. Natl. Cancer Inst.* **50**:1051–1053.

Giovanella, B. C., Stehlin, J. S., and Williams, L. J. (1974). Heterotransplantation of human malignant tumors in nude, thymusless mice. II. Malignant tumors induced by the injection of cell cultures derived from human solid tumors. *J. Natl. Cancer Inst.* **52**:921–927.

Goding, J. W., and Burns, G. F. (1981). Monoclonal antibody OKT-9 recognizes the receptor for transferrin on human acute lymphocytic leukemia cells. *J. Immunol.* **127**:1256–1258.

Goldenberg, D. M., DeLand, F., Kim, E., Bennett, S., Primus, F. J., Van Nagell, J. R., Estes, N., DeSimone, P., and Rayburn, P. (1978). Use of radiolabeled antibodies to carcinoembryonic antigen for the detection and localization of diverse cancers by external photoscanning. *N. Engl. J. Med.* **298**:1384–1388.

Goldenberg, D. M., Kim, E., and DeLand, F. (1980). Clinical studies on the radioimmunodetection of tumors containing alpha fetoprotein. *Cancer* **45**:2500–2512.

Hart, I. R., and Fidler, I. J. (1981). The implications of tumor heterogeneity for studies on the biology and therapy of cancer metastases. *Biochem. Biophys. Acta.* **651**:37–50.

Hoffer, P. (1980). Gallium: mechanisms. *J. Nucl. Med.* **21**:282–285.

Karush, F. (1957). The interaction of purified anti β-galactoside antibody with haptens. *J. Am. Chem. Soc.* **79**:3380–3384.

Kasai, M., Saxton, R. E., Holmes, B. C., Burk, M. W., and Morton, D. L. (1981). Membrane antigens detected on human lung carcinoma cells by hybridoma monoclonal antibody. *J. Surg. Res.* **30**:403–408.

Kennett, R. H., and McKearn, T. J., and Bechtol, K. B. (1980). *Monoclonal Antibodies*. New York: Plenum Press.

Kerbel, R. S. (1979). Selection of high immunogenicity in drug resistant sublines of murine lymphomas demonstrated by plague assay. *Nature* **280**:358–360.

Khaw, B. A., Beller, G. A., Haber, E., and Smith, T. W. (1976). Localization of cardiac myosin-specific antibody in myocardial infarction. *J. Clin. Invest.* **58:**439–446.

Khaw, B. A., Fallon, J. T., Strauss, H. W., and Haber, E. (1980). Myocardial infarct imaging of antibodies to canine cardiac myosin with indium-111 diethylenetriamine pentaacetic acid. *Science* **209:**295–297.

Killion, J. J., and Kollmorgen, G. M. (1976). Isolation of immunogenic tumor cells by cell affinity chromatography. *Nature* **259:**674–676.

Köhler, G., and Milstein, C. (1975). Continuous cultures of fused cells secreting antibody of predefined specificity. *Nature* **256:**495–497.

Krejcarek, G. E., and Tucker, K. L. (1977). Covalent attachment of chelating groups to macromolecules. *Biochem. Biophys. Res. Comm.* **77:**581–585.

Kyriazis, A. A., and Kyriazis, A. P. (1980). Preferential sites of growth of human tumors in nude mice following subcutaneous transplantation. *Cancer Res.* **40:**4509–4511.

Kyriazis, A. P., Pesce, A. J., DiPersio, L., Stinnett, D., and Michael, J. G. (1978). Metastatic behavior of human tumors growing in athymic mice. *Fed. Proc.* **37:**680.

Kyriazis, A. P., DiPersio, L., Michael, G. J., Pesce, A. J., and Stinnett, J. D. (1978). Growth patterns and metastatic behavior of human tumors growing in athymic mice. *Cancer Res.* **38:**3186–3190.

Lampson, L. A. (1980). Binding inhibition using biosynthetically labeled monoclonal antibody. In *Monoclonal Antibodies,* Kennett, R. H., and McKearn, T. J. eds. New York: Plenum Press, p. 398.

Larson, S. M. (1978). Mechanisms of localization of gallium-67 in tumors. *Seminars Nucl. Med.* **8:**193–203.

Larson, S. M., Rasey, J. S., Allen, D. R., Nelson, N. J., Grunbaum, Z., Harp, G. D., and Williams, D. L. (1980). Common pathway for tumor cell uptake of gallium-67 and iron-59 via a transferrin receptor. *J. Natl. Can Inst.* **64:**41–53.

Levine, G., Ballou, B., Reiland, J., Solter, D., Gumerman, L., and Hakala, T. (1980). Localization of ^{131}I labeled tumor specific monoclonal antibody in the tumor bearing BALB/c mouse. *J. Nucl. Med.* **21:**570–573.

Levy, R., and Dilley, J. (1978). *Proc. Natl. Acad. Sci. USA* **75:**2411–2415.

Mach, J. P., Carrel, S., Merenda, C., Sordat, B., and Cerrottini, J. C. (1974). *In-vivo* localization of radiolabeled antibodies to carcinoembryonic antigen in human colon carcinoma grafted into nude mice. *Nature* **248:**704–706.

Mach, J. P., Forni, M., Ritschard, J., Buchegger, F., Carrel, S., Widgram, S., Donath, A., and Alberto, P. (1980). Use and limitations of radiolabeled anti-CEA antibodies and their fragments for photoscanning detection of human colorectal carcinomas. *Oncodevelopmental Biol. Med.* **1:**49–69.

Mach, J. P., Carrel, S., Forni, M., Ritschard, J., Donath, A., and Alberto, P. (1980). Tumor localization of radiolabeled antibodies against carcinoembryonic antigens in patients with carcinoma. A critical evauation. *N. Engl. J. Med.* **303:**5–10.

Mason, D. W., and Williams, A. F. (1980). The kinetics of antibody binding to membrane antigens in solution and at the cell surface. *Biochem. J.* **187:**1–20.

McGee, J., Woods, J. C., Ashall, F., Bramwell, M. E., and Harris, H. (1982). A new marker for human cells. 2. Immunohistochemical detection of the Ca antigen in human tissues with Ca1 antibody. *Lancet* **3**(July):7–10.

Milstein, C. (1980). Monoclonal antibodies. *Sci. Am.* **243:**66–77.

Mitchell, K. F., Fuhrer, J. P., Steplewski, Z., and Koprowski, H. (1980). Biochemical characterization of human melanoma cell surfaces: Dissection with monoclonal antibodies. *Proc. Natl. Acad. Sci. USA* **77**:7287–7291.

Moshakis, V., McIlhinney, R. A. J., Raghavan, D., and Neville, A. M. (1981). Monoclonal antibodies to detect human tumors: An experimental approach. *J. Clin. Pathol.* **34**:314–319.

Moshakis, V., McIlhinney, R. A. J., Raghaven, D., and Neville, A. M. (1981). Localization of human tumor xenografts after I.V. administration of radiolabeled monoclonal antibodies. *Brit. J. Cancer* **44**:91–99.

Nahmias, C., Ikeno, C., and Coastes, G. (1981). The transcapillary escape rate of transferrin. *Microvascular Res.* **21**:128–132.

Nisonoff, A., Harper, J. E., and Spring, S. R. (1975). *The Antibody Molecule.* New York: Academic Press.

Noujaim, A. A., Lentle, B. C., Hill, J. R., Terner, U. K., and Wong, H. (1979). On the role of transferrin in the uptake of gallium by tumor cells. *Int. J. Nucl. Med. Biol.* **6**:193–199.

Old, L. J. (1981). Cancer immunology: The search for specificity. *Cancer Res.* **41**:361–375.

Olsson, L., and Kaplan, H. S. (1980). *Proc. Natl. Acad. Sci. USA* **77**:5429–5431.

Omary, M. D., and Trowbridge, I. S. (1981). Biosynthesis of the human transferrin receptor in cultured cells. *J. Biol. Chem.* **256**:12888–12892.

Paik, C. H., Herman, D. E., Eckelman, W. C., and Reba, R. C. (1980). Synthesis, plasma clearance and *in-vitro* stability of protein containing a conjugated indium-111 chelate *J. Radioanalyt. Chem.* **57**:553–564.

Paik, C. H., Murphy, P. R., Eckelman, W. C., Valkert, W. A., and Reba, R. C. (1982). The conjugation of metal chelates to antibodies. *J. Nucl. Med.* **23**:37.

Pantelouris, E. M. (1968). Absence of a thymus in a mouse mutant. *Nature* **217**:370–371.

Pettit, W. A., DeLand, F. H., Bennett, S. J., and Goldenberg, D. M. (1980). Improved protein labeling by stannous tartrate reduction of pertechnetate. *J. Nucl. Med.* **21**:59–62.

Porter, R. R. (1959). The hydrolysis of gamma globulin and antibodies with crystalline papain. *Biochem. J.* **73**:119–126.

Povlsen, C. O. (1978). Growth of human tumors in nude mice and chemotherapy studies in this model. In *Proceedings of the Symposium on the Use of Athymic (Nude) Mice in Cancer Research,* Houchens, D. P., and Ovejera, A. A., eds. Stuttgart: Gustav Fischer.

Prehn, R. T. (1970). Analysis of antigenic heterogeneity within individual 3-methylcholanthrene-induced mouse sarcomas. *J. Natl. Cancer Inst.* **45**:1039–1043.

Pressman, D., and Korngold, L. (1953). The *in-vivo* localization of anti-Wagner osteogenic sarcoma antibodies. *Cancer* **6**:619–626.

Pressman, D., Day, E. D., and Blau, M. (1957). The use of paired antibody labeling in the determination of tumor localizing antibodies. *Cancer Res.* **17**:845–850.

Primus, F. J., Wang, R. H., Goldenberg, D. M., and Hansen, H. J. (1973). Localization of human GW-39 tumors in hamsters by radiolabeled heterospecific antibody to carcinoembryonic antigen. *Cancer Res.* **33**:2977–2987.

Putnam, F. W. (1977). In *The Plasma Proteins,* Vol. III, 2nd ed., Putnam, F. W., ed. New York: Academic Press.

Rauscher, F. J. (1962). A virus-induced disease of mice characterized by erythrocytopoiesis and lymphoid leukemia. *J. Natl. Cancer Inst.* **29:**515–543.

Reid, L. M., Holland, J., Jones, C., Wolf, B., Niwayama, G., Williams, R., Kaplan, N. O., and Sato, J. (1978). Some of the variables affecting the success of transplantation of human tumors into the athymic (nude) mouse. In *Proceedings of the Symposium on the Use of Athymic (Nude) Mice in Cancer Research,* Houchens, D. P., and Ovejera, A. A., eds. Stuttgart: Gustav Fischer.

Rygaard, J. (1969). Immunobiology of the mouse mutant "nude." Preliminary investigations. *Acta Pathol. Microbiol. Scand.* **77:**764–772.

Rygaard, J. (1978). The nude mouse–mouse or test tube. In *Proceedings of the Symposium on the Use of Athymic (Nude) Mice in Cancer Research,* Houchens, D. P., and Ovejera, A. A., eds. Stuttgart: Gustav Fischer, p. 1.

Rygaard, J., and Povlsen, C. O. (1969). Heterotransplantation of a human malignant tumor to the mouse mutant "nude." *Acta Pathol. Microbiol. Scand.* **77:**758–760.

Sands, H., Jones, P. L., Neacy, W., Camin, L. L., Gallagher, B. M. (1983). Imaging of small experimental tumors in the subrenal capsule—a model for imaging human metastases (submitted).

Sands, H., Jones, P. L., Neacy, W., Camin, L. L., Gallagher, B. M. (1983). Unpublished results.

Scatchard, D. (1949). The attraction of proteins for small molecules and ions. *Ann. N.Y. Acad. Sci.* **51:**660–672.

Schwartz, M., Lancet, D., Mozes, E., and Sela, M. (1978). Affinity and ayidity of antibodies to the random polymer (T.G.) -A-L. and related ordered synthetic polypeptide. *Immunochemistry* **15:**477–481.

Sharkey, F. E., and Fogh, J. (1979). Metastasis of human tumors in athymic nude mice. *Int. J. Cancer* **24:**733–738.

Sheinberg, D. A., and Strand, M. (1982). Leukemic cell targeting and therapy by monoclonal antibody in a mouse model system. *Cancer Res.* **42:**44–49.

Sheinberg, D. A., Strand, M., and Gansow, O. A. (1982). Tumor imaging with radioactive metal chelates conjugated to monoclonal antibodies. *Science* **215:**1511–1513.

Shimosato, Y., Kameya, T., Koide, T., Oboshi, S., Seido, T., Hayata, Y. Tsuji, K., and and Nomura, T. (1974). Morphological study of human lung cancers transplanted in nude mice. *Proc. Jap. Cancer Assoc.* **33:**23–29.

Shimosato, Y., Kameya, T., Nagai, K., Hirohashi, S., Koide, T., Hayoshi, H., and Namura, T. (1976). Transplantation of human tumors in nude mice. *J. Natl. Cancer Inst.* **56:**1251–1255.

Shindelman, J. E., Ortmeyer, A. E., and Sussman, H. H. (1981). Demonstration of the transferrin receptor in human breast cancer tissue. Potential marker for identifying dividing cells. *Int. J. Cancer* **27:**329–334.

Sibora, K., and Wright, R. (1981). Human monoclonal antibodies to lung cancer antigens. *Brit. J. Cancer* **43:**693–703.

Sips, R. (1948). On the structure of a catalyst surface. *J. Chem. Phys.* **16:**490–493.

Sordat, B., Fritsche, R., Mach, J. P., Carrel, S., Ozzello, L., Cerottini, J. C. (1974). Morphological and functional evaluation of human solid tumors serially transplanted in nude mice. In *Proceedings of the First International Workshop on Nude Mice,* Rygaard, J., and Povlsen, C. O., eds. Stuttgart: Gustav Fischer.

Smith, T. W., Curd, J. F., and Haber, E. (1971). Rapid reversal of digoxin toxicity by Fab fragments from purified digoxin specific antibody. *Circulation* **44:**11–41.

Smith, T. W., Haber, E., Yeatman, L., and Butler, V. P. (1976). Reversal of advanced digoxin intoxication with Fab fragments of digoxin-specific antibodies. *N. Eng. J. Med.* **294**:797–800.

Smith, T. W., Butler, V. P., and Haber, E. (1977). Cardiac glycoside-specific antibodies in the treatment of digitalis intoxication. In *Antibodies in Human Diagnosis and Therapy,* Haber, E., and Krause, R. M., eds. New York: Raven Press, pp. 365–388.

Soloman, A. (1980). In *Cancer Markers: Diagnostic and Developmental Significance,* Sell, S., ed. Clifton, N.J.: The Humana Press.

Spiegelberg, H. L., and Weigle, W. O. (1965). The catabolism of homologous and heterologous 7S gamma globulin fragments. *J. Exp. Med.* **121**:323–338.

Tax, A., and Manson, L. (1981). Monoclonal antibodies against antigens displayed on a progressively growing mammary tumor. *Proc. Natl. Acad. Sci. USA* **78**:529–533.

Van Nagell, J. R., Kim, E., Casper, S., Primus, F. J., Bennett, S., DeLand, F. H., and Goldenberg, D. M. (1980). Radioimmunodetection of primary and metastatic ovarian cancer using radiolabeled antibodies to carcinoembryonic antigen. *Cancer Res.* **40**:502–506.

Waldmann, T. A., and Strober, W. (1969). Metabolism of immunoglobulins. *Prog. Allergy* **13**:1–110.

Waldman, T. A, Blaese, R. M., and Strober, W. (1970). In *Plasma Protein Metabolism,* Rothschild, M. A., and Waldmann, T., eds. New York: Academic Press.

Wilson, E. M., French, F. S., and Petrusz, P. (1982). Transferrin in the rat prostate Dunning tumor. *Cancer Res.* **42**:243–251.

Witebsky, E. (1929). Disponibilität und Spezifität alkohollöslicher strukturen von Organen und bosartigen Geschwülsten. *Z. Immunitaetsforsch.* **62**:35–73.

Wong, H., Terner, U. K., English, D., Noujaim, A. A., Lentle, B. C., and Hill, J. R. (1980). The role of transferrin in the *in-vivo* uptake of gallium-67 in a canine tumor. *Int. J. Nucl. Med. Biol.* **7**:9–16.

Yeh, M. Y., Hellström, I., Brown, J. P., Warner, G. A., Hansen, J. A., and Hellstöm, K. E. (1979). Cell surface antigens of human melanoma identified by monoclonal antibody. *Proc. Natl. Acad. Sci. USA* **76**:2927–2931.

Yelton, D. E., and Scharff, M. E. (1981). Moncolonal antibodies: A powerful new tool in biology and medicine. *Ann. Rev. Biochem.* **50**:657–680.

CHAPTER 4

Small Animal Oncological Models for Screening Diagnostic Radiotracers

LEONARD I. WIEBE

INTRODUCTION

Oncological models that are used to study diagnostic or therapeutic aspects of neoplastic disease can be assigned to either of two major groups. The *in vitro* models include cultured cells, cell suspensions, and isolated cell systems; the *in vivo* models are based on the growth of neoplastic tissue in an appropriate host. Both systems are characterized by a number of advantages and disadvantages, some of which they share and others of which apply uniquely to them. The *in vitro* models, for example, are less sensitive to the acute toxicity of test chemicals but are completely independent of influencing factors derived from the tumor host and may be sensitive to minute changes in their artificial environment. On the other hand, the *in vivo* models are more likely to give a representative response in terms of overall effect but may poorly model drug bioavailability, pathological stresses, or growth characteristics. *In vitro* and *in vivo* systems both have the need for periodic reidentification to ensure that bacterial contamination and cell mutation, and indeed invasion by other neoplastic cells, has not occurred. The detective work of Walter Nelson-Rees, which was recently drawn to public attention (Gold, 1981), is an example of the latter problem. For almost a decade Nelson-Rees has been tracking down the identity of neoplastic cells in various systems. He has succeeded in identifying many cell lines, cells that had been labeled otherwise, as Hela cells. He has shown HBT-3 cultures, HEK cultures, and many others to be chromasomally similar to the cervical cancer cells harvested from Henrietta Lacks in 1951.

In vitro and *in vivo* oncological models have both been used extensively for the screening of diagnostic radiotracers. The *in vitro* systems have gained some additional prominence during the past few years, particularly for gallium and for steroid receptor studies. Quantitatively, the *in vitro* models represent only a small fraction of this literature, perhaps because these models are not construed as the ultimate preclinical test.

In vivo oncological models, that is, tumor-bearing laboratory animals, are the subject of this treatise. An attempt has been made to reiterate, unify, and extend the substance of earlier reviews. Table 4-5 through 4-10 provide a comprehensive, but not exhaustive, listing of animal tumor models, radioactive test substances, and literature citations for the period 1976–1981, inclusively. This compilation supplements the most recently published review of the subject (Haynie et al., 1976).

THE NEED FOR ANIMAL TUMOR MODELS

The question might well be asked, "Why use animal models for cancer studies involving diagnostic radiotracers?" The question is not raised irresponsibly. It is commonly felt that most (if not all) animal models have one or more major flaws that preclude an unequivocable extension of the findings to the human condition. Furthermore, a number of physicians and patients alike would willingly participate in a well-rationalized diagnostic long-shot. The opinions expressed by two authors can be paraphrased to explain why animal tumor models should be and are being used. It has been stated that cancer is unlike other human diseases in that the general biological features of human cancers are faithfully represented in spontaneous cancers of small mammals used in laboratory investigations (Hewitt, 1978). In other words, we can expect the model to be reasonably representative if it is chosen with care. It has also been stated that animal models are useful in the development of tumor-seeking agents and are an essential prerequisite to use of the drug in humans (Halpern, 1977). Such preclinical testing of any diagnostic or therapeutic protocol has, in fact, been the underlying principle of the ethics of clinical investigation in modern medicine.

There are a number of similarities among human and animal cancers. The potentials for carcinogenesis by physical, chemical, and biological agents, for example, are similar. Their general biological features, such as blood supply, cell surface properties and molecular structure of the cell membranes, and deviant metabolic patterns are frequently shared. Certainly the mechanisms of growth, the growth and spreading patterns, and the agencies affecting growth and restraint of growth are shared.

CANCER

Just as growth can be considered the common characteristic of living systems, orderly or regulated growth can be considered the hallmark of normal multicellular organisms. The influence of pathological processes on normal growth may result in the alteration of such growth, and consequently, growth may become characterized by malformation, excessive or diminished growth, disturbance in cellular differentiation, disturbance in organ growth, or autonomous growth.

Excessive growth may involve increased cell size (hypertrophy) or increased cell numbers (hyperplasia). Physiologic hyperplasia is distinguishable from neoplastic hyperplasia (neoplasia) in that the latter is due to a heritable cellular abnormality, whereas the former will subside when the external stimulus is removed.

The words *tumor* and *cancer* are frequently used in the context of *neoplasm* and *malignant neoplasm,* although the terms are not strictly synonymous. Neoplastic growth is a more restrictive term than *tumor* or *cancer,* in that it implies

1. Excessive, uncontrolled growth, which continues after removal of the original stimulus.
2. The formation of cells that are permanently altered.
3. The development of cells in which differentiation is decreased.
4. An altered nuclear-cytoplasmic relationship.
5. Cellular dysfunction.
6. Encapsulated, noninvasive growth that may be benign.
7. Anaplastic, infiltrative, rapid, metastatic growth that may be malignant.

A large number of excellent reference texts for general cancer biology have been published. The titles *Biology of Cancer* (Ambrosie and Roe, 1975), *Neoplasms—Comparative Pathology of Growth in Animals* (Kaiser, 1978), *Cancer Pathophysiology, Etiology and Management* (Kruse et al., 1979) and *Clinical Oncology* (Horton and Hill, 1979) are highly recommended.

CLASSIFICATION OF NEOPLASMS

Neoplasms may be classified in terms of their behavioristic or histeogenetic characteristics. The behavior classification lists neoplasms as being either benign or malignant. Most, if not all, tumor model studies with diagnostic radiotracers are oriented toward the study of malignant neoplasms. Several of the more frequently used histeogenetic classes with examples of malignant and benign neoplasms (Louis 1978) include

1. Epithelial origin: papillomas (benign) or adenocarcinomas (malignant).
2. Connective tissue: fibromas and osteomas (benign) or fibrosarcomas and osteosarcomas (malignant).
3. Muscle: rhabdomyoma (benign) or rhabdomyosarcoma (malignant).

4. Nervous tissue: ganglioneuroma (benign) or neuroblastoma and glioma (malignant).
5. Reticuloendothelial: Hodgkin's lymphoma (malignant).
6. Others: malignant melanomas, malignant teratomas.

GROWTH AND SPREAD OF NEOPLASMS

The malignant transformation of cells has been characterized, but the underlying causative factors and growth mechanisms remain the subject of intense study. The main features of malignant transformation involve a sequence of pleomorphism, increased growth, invasive growth, anaplasia, and metastasis. Pleomorphism includes a number of phenomena that result in the cell's escape from the normal control mechanisms. This is associated with changes in cell surface shape and adhesiveness, increased fucose-containing glycoprotein, the development of tumor-specific glycoprotein antigens, and the suppression of plasma membrane glycolipids. In general, there is an increased growth rate, which is associated with abnormally invasive growth. The mechanisms responsible for this invasiveness include mechanical pressure, tumor angiogenesis, reduced cellular adhesiveness, the release of toxins, phagocytosis, and immunological factors. The outgrowths are characterized by cellular inferiority, anaplasia, and dysfunction.

The ability of certain neoplasms to spread (metastasize) serves as a basis for classifying neoplasms as benign or malignant. The ability to metastasize, although dependent on a number of characteristics of the neoplasms itself, is also dependent on the ability of the host to provide secondary, remote sites for unrestricted growth of the isolated cells. Vectors of metastases are the lymphatic system, the bloodstream, and certain body cavities. The lymphatic vector is a common medium for the spread of carcinomas. It includes lymph nodes, lymphatic vessels, and perineural lymphatics, which permit metastasis by the mechanisms of permeation (solid cords) or embolization (cell clumps). Hematogenous metastasis is common for sarcomas and some carcinomas. The mechanism involves invasion of small venules, and as a result is responsible mainly for pulmonary and hepatic metastatic growth. Carcinomas of the stomach, colon, and ovary metastasize via the peritoneal cavity; carcinomas of lung, breast, and esophagus metastasize via pleural and pericardial cavities; and choroid plexus tumors, ependymomas, and medulloblastomas metastasize via the cerebrospinal cavity.

ANIMAL TUMOR MODELS

The selection of an appropriate *in vivo* oncological model must be based on criteria that define both the malignant neoplasm and the host or carrier species and strain. The pragmatic aspects are at least as important as scientific consideration in many instances. Major constraints include

1. Size of host.
2. Availability and cost of the host.
3. Ease of handling the host animal, including ease of administration of the radiotracer via the appropriate route.
4. Similarity of the model organ to the human tumor-bearing organ being modeled.
5. Host and human rhythms.

In addition to the basic considerations governed by practicality, the host–tumor relationship must parallel that of the clinical syndrome in terms of organ function–dysfunction, humoral influences, and even fundamental anatomical features. Modeling for circadian rhythms, free-running rhythms, diurnal rhythms, and the dominating synchronizing forces (light, food, drugs) has been found to be an integral part of host selection. Several current reference books (Altman and Katz, 1979; Mitruka, 1976) are available to aid both novice and experienced researchers. In general cancer research the most frequently utilized hosts are mice, followed at some distance by rats and other species. Hewitt (Hewitt, 1978) reported that in 1975 the relative usage of animal tumor hosts was 63% for mice, followed by 32% for rats, 3.4% for hamsters, and 1.5% for guinea pigs. The remainder (0.5%) included rabbits, dogs, bovine, and fowl.

To some extent, the choice of an appropriate neoplasm is limited by the selection of host animal. However, there is a large choice of well-documented tumors available for a given species (Hewitt, 1978; Schabel, 1975; Roberts and Tait, 1977). This effectively eliminates this constraint, and the development of immunologically nude animals now permits a broad range of interspecies transplants (Helson et al., 1975). The selection of an appropriate tumor for the host species chosen for the research should be based on the following criteria:

1. The ease of tumor generation in new hosts.
2. The potential of the tumor for *in vitro* replication and study.
3. The influence of the tumor on systemic physiological and metabolic processes in the host.
4. Anatomical, histological, physiological, and biochemical similarities between the animal tumor and the human tumor being modeled.
5. Whether the tumor is encapsulated or invasive. This is important particularly when the experimental protocol requires precise definition (including surgical removal) of the tumor.
6. Artifactual tumor immunogenicity.
7. Tumor growth rate.
8. Tumor chronobiology.

Tumor growth rate is important not only from the viewpoint of uptake studies and metabolic requirements, but also with respect to time required to develop the model. Thirty to thirty-five doubling events are required to bring the tumor to a size that can be diagnosed (1 cm^3), but another half-dozen divisions will increase tumor size to the point where the host will succumb. Therefore, a tumor that reaches a 1 cm^3 size within a week, or two weeks, will limit the life expectancy of the host to only a few days thereafter. Tumor growth and the number of cells in the tumor are expressed as a function of the number of doubling events in Fig. 4-1. It is clear that careful selection of a tumor model is required if problems with tumor supply logistics and experimental protocols are to be avoided.

Although animal neoplasms are classified in terms of their behavioral and histeogenetic characteristics, as mentioned previously, the animal tumor model is frequently selected on the basis of method of tumor propagation or development. Thus tumors may also be classified as spontaneous, induced, transmissible, or transplantable.

Spontaneous tumors are used only rarely in studies with small laboratory rodents, but they are heavily relied on for large-animal investigations. These tumors have a low incidence, and the researcher has no control over the anatomical site at which they will develop, the type of tumor that will appear, or the host age at the time of tumor development. The advantages of these tumors may include model suitability with respect to anatomical site, blood supply, and induced systemic responses. Examples of the occurrence of spontaneous tumors in domestic dogs and cats (Priester and Mantel, 1971) are given in Tables 4-1 and 4-2.

Induced and transplantable tumors are used mort frequently in models that have small rodents as hosts. Induced tumors have a variable latent

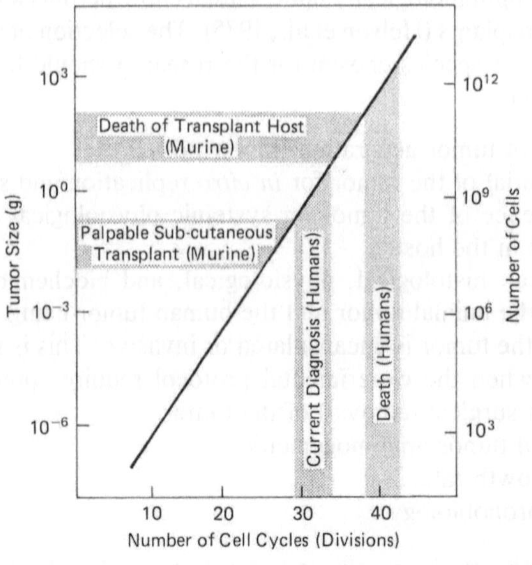

FIG. 4-1. Tumor growth as a function of cell division in an ideal system. In real systems, daughter death, irregularities in division, and a number of environmental conditions induce irregularities among these relationships.

TABLE 4-1. Spontaneous Canine Neoplasms

Anatomical Site	*Percentage of Neoplasms*	*Percentage of Malignancies*
Skin	28	20
Mammary	12	46
Digestive	10	33
Blood-lymph	7	95
GU system	7	37
Oral	6	49
Bone	5	81

period that is characteristic of the tumor and the inducer. The researcher has little or no control over anatomic site or tumor type, and, in fact, the tumor(s) may develop where no spontaneous tumor is known or readily available. The advantages of induced tumors over spontaneous tumors lie primarily in their high incidence of occurrence and in their earlier development. As with spontaneous tumors, they may model anatomical site, blood supply, and physiological and metabolic stresses of the human tumor.

Transplantable tumors are utilized more frequently than induced or spontaneous tumors, perhaps because they can be generated on a definite schedule, and because they do not require initial histological identification (this having been done by the tumor developer–supplier). Many laboratories that conduct screening (of radiotracers) studies do not have the facilities or the expertise to undertake such controls, and because of cost and inconvenience, prefer to utilize well-established tumor lines. The potential for loss of identity by these tumor lines must be recognized, and periodic recharacterization or replacement from a supply source should be undertaken.

Transmissible and transplantable tumors are characterized in general terms by their abnormal anatomical location and blood supply and the absence of systemic effects in the host. The researcher has control over the number of hosts, host age, and tumor size; anatomical location; and histol-

TABLE 4-2. Spontaneous Feline Neoplasms

Anatomical Site	*Percentage of Neoplasms*	*Percentage of Malignancies*
Blood-lymph	35	98
Skin	18	41
Oral	8	65
Digestive	7	71
Mammary	5	78
Respiratory	5	78
Bone	4	47

TABLE 4-3. Some Transplantable Animal Tumors that Undergo Metastasis

Host Species	*Tumor (host strain)*
Mice	Cloudman melanoma (DBA mice)
	Ehrlich carcinoma (BALB/c mice)
	Sarcoma T241 Lewis (C57BL mice)
	Sarcoma S37 (BALB/c and DBA2 mice)
Rats	Walker carcinosarcoma
	Flexner–Jobling carcinoma
	Jensen sarcoma
	Yoshida sarcoma
Hamsters	Amelanotic melanoma (Fortner)
	lymphomas (ML and NML)
Rabbits	Brown–Pearce carcinoma
	V2 carcinoma

ogy. Allografts are used most frequently. They are serially transplantable, require only periodic check for histologic stability, and may be passed on to a new host via IV, IM, IP, and/or sub-Q implantation routes. There is H-2 histocompatibility within species and strain. Xenografts are less suitable for serial transplantation, although a number of cell lines now exist that are not self-limiting in this regard. However, early anaplastic transformation is common in many of these lines. The host must be immunologically suppressed or immunodeficient, or one may use immunodeficient (privileged) sites in a normal host. The immunodeficient hosts require special care to protect them from invasion by other diseases. Although xenografts of human tumors for use in screening studies are reported with increasing frequency, a number of practical difficulties, including requirements for excellent host housing and for tumor biology expertise, serve as effective deterrents to their widespread use.

The ability of a tumor to metastasize is a characteristic that may be desirable when modeling for human neoplastic disease. Examples of transplantable metastasizing animal tumors that have been used for screening diagnostic radiotracers in the most commonly used hosts are given in Table 4-3.

SELECTION OF ANIMAL TUMOR MODELS BASED ON MECHANISM OF UPTAKE OF THE RADIOTRACER

The biochemical and immunological characteristics of the tumor model, as well as tumor blood flow, lymphatic supply, and pleomorphic changes are the most common bases for radiotracer design. It is obvious that both the human neoplasm and its animal model should share the common trait that has been selected for exploitation. An incomplete list of such exploitable characteristics/mechanisms includes both physiological, pharmacological, and biochemical properties:

1. Increased permeability of capillary beds to macromolecules.
2. Delayed formation of lymphatic vessels.
3. Increased macrophage activity and necrosis.
4. Increased cell membrane permeability.
5. Antigenicity–immunological properties–fetal macromolecules.
6. Receptors–transport sites.
7. Hypoxia/anoxia.
8. Blood flow and blood vessel formation.
9. Increased glycolysis; decreased gluconeogensis.
10. Enhanced anabolic, decreased catabolic activity on DNA pathway.
11. Transcription and polymerase changes.
12. Altered isoenzymes.
13. Hormones.

ANIMAL TUMOR MODELS IN THE LITERATURE OF RADIOTRACERS AND RADIOPHARMACEUTICALS

From 1976 through 1981 there were more than 160 published abstracts and full papers describing radiotracer screening procedures using animal tumor models. The mouse remains the most frequently utilized host, but there appears to be a trend toward more frequent use of larger animals, particularly the rat and the dog. Surprisingly, the use of hamsters has declined, although many tumor biologists feel that this species is particularly well suited to model for pancreatic, bladder, and respiratory neoplasms. A comparison of host species used before 1976, with those used in the 1976–1981 period is provided in Table 4-4. In approximately 10% of the literature cited, the host species was not named, although in some cases it could be deduced from the tumor description provided.

TABLE 4-4. Hosts for Tumor Models Used to Test Diagnostic Radiotracers

Host Species	*1976–1981 Percent of Total*[a]	*Haynie et al., 1976 Percent of Total*
Mouse	47	54
Rat	36	23
Dog	7	4.5
Hamster	6	14
Rabbit	4	4.5

[a] Based on a search of 12 journals from 1976 to 1981 inclusively.

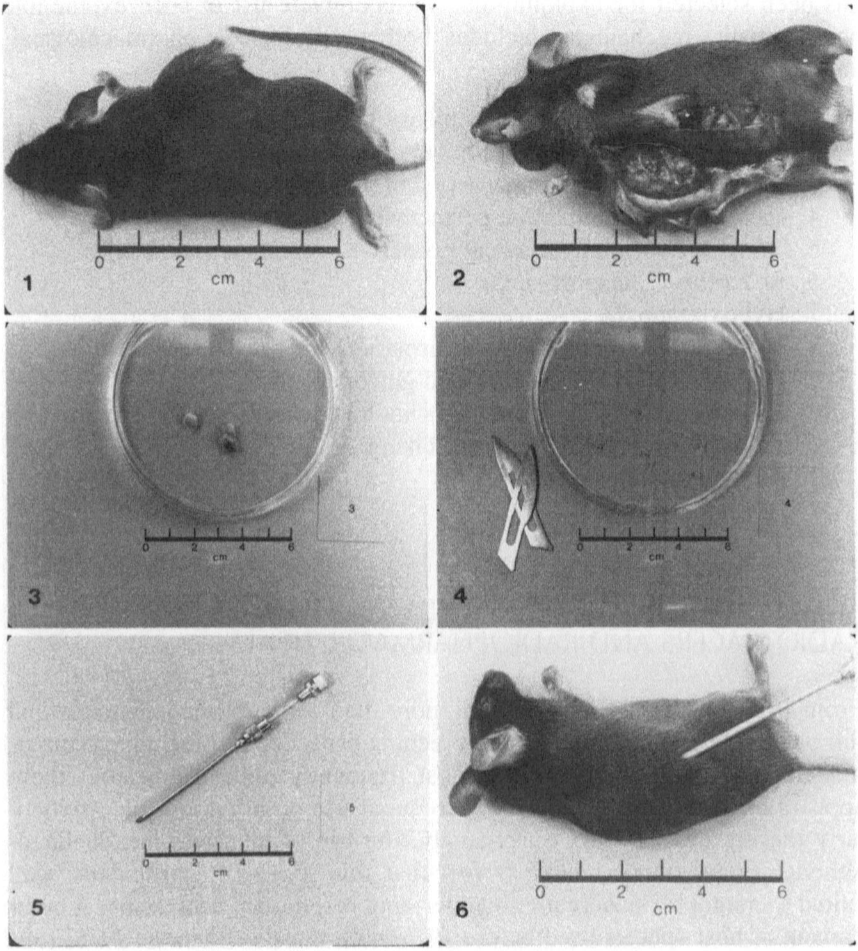

FIG. 4-2. Propagation of the Lewis Lung carcinoma in BDF_1 mice. (1) A mature tumor. (2) The surgically exposed tumor. (3) The excised tumor in cross section, showing an outer region of viable cells suitable for transplant. (4) The tumor minced to 1-mm^3-sized plugs in saline. (5) A small plug loaded into a trochar and cannula, ready for insertion. (6) Subcutaneous insertion of the carcinoma into a lightly anesthetized mouse. (Photograph courtesy of Yip Wan Lee.)

The Lewis Lung carcinoma, a malignant neoplasm propagated in BDF_1 mice, is of particular interest because it has been adopted by the Division of Cancer Treatment of the U.S. National Cancer Institute for screening chemotherapeutic agents (Hewitt, 1978). This tumor line was developed by Lewis in 1951 by transplanting a spontaneous epidermoid carcinoma from the lung of a C57BL mouse into BDF_1 (C57BL/6 × DBA/2) inbred mice (Sugiura and Stock, 1955). The tumor is characterized by rapid growth (to >1 cm^3 in 14 days), a high subcutaneous transplant "take" rate (virtually

100%), a very low occurrence of spontaneous regression (about 4%), and the early development of lung and kidney metastases. The tumor may be transplanted by intravenous, intramuscular, or subcutaneous injection of cell suspensions, or by subdermal implants, depending on the requirements of the research protocol. The implant technique depicted in Fig. 4-2 is a rapid and effective method to propagate the tumor line for use in screening radiotracers.

Examples of animal tumor models used, listing tumor name and description, host species and strain if known, and radiotracers tested, are given in Tables 4-5 through 4-10.

TABLE 4-5. Uptake of Radiotracer by Mouse Tumors, 1976–1981

Tumor	*Host*	*Radiotracer*	*Reference*
Mammary adenocarcinoma MMI (slow) MMII (intermediate) KHJJ (fast)	mice	[*I]bleomycins (4) [*I]HSA [*I]fibrinogen [^{67}Ga]citrate	DeNardo et al., 1976
B-16 melanoma	mice	[^{11}C]*N*-methyl-1,4-diamino-butane	Welch et al., 1976
Human neuroblastoma	nude mice	[^{13}N]BCNU	Petit et al., 1976
Mammary adenocarcinoma MMI MMII KHJJ	BALB/c	[*I]fibrinogen	Wortman et al., 1976
Mammary carcinoma	BALB/c	[^{125}I]antiMMEC	Ceriani et al., 1976
Ridgeway osteogenic sarcoma	AKD2-F1	[^{67}Cu,^{111}In,^{57}Co] -bleomycin -citrate	Hall et al., 1976
Adenocarcinoma	mice	[^{64}Cu]thiosemicarbazone	Lieberman et al., 1976
Fibrosarcoma	mice	[^{64}Cu]thiosemicarbazone	Lieberman et al., 1976
B-16 melanoma	mice	[^{11}C]1-aminocyclopentane-carboxylic acid	Hayes et al., 1976
Sarcoma-180	ICR	[^{153}Sm]citrate	Friedman et al., 1976
Lewis Lung	C57BL/6	[^{124}Sb]chloride	Robbins et al., 1976
Glioma	C57BL/6	[^{124}Sb]chloride	Robbins et al., 1976
Ridgeway osteosarcoma	AKR	[^{124}Sb]chloride	Robbins et al., 1976

TABLE 4-5. (Continued)

Tumor	*Host*	*Radiotracer*	*Reference*
NCTC 2472 (A68) fibrosarcoma	C3H/He	methionine-methyl-^{11}C	Comar et al., 1976
Ehrlich ascites	mice	[^{125}I]fibrinogen [^{203}Hg]thymidine	Zimmerman and Schmutz, 1976
B-16	mice	[^{11}C]putrescine -spermine -spermidine	Welch et al.
KHJJ adenocarcinoma	BALB/c	[^{131}I]bleomycin [^{67}Ga]citrate [^{99m}Tc]bleomycin [^{57}Co]bleomycin [^{111}In]bleomycin	Krohn et al., 1977
Ehrlich	CF-1	[^{59}Fe]chloride [^{59}Fe]metalocene carboxylic acid [^{103}Ru]chloride [^{103}Ru]metalocene carboxylic acid [^{99m}Tc]Fe-ascorbate -Fe-ascorbate-DTPA [^{67}Ga]citrate	Wenzel et al., 1977
Sarcoma-180	ICR	[^{57}Co]bleomycin	Zalutsky et al., 1977
L1210 leukemia FU responsive FU resistant	mice	[5-^{18}F]U	Wolf et al., 1977
Rideway osteogenic sarcoma	AKD-2	[^{57}Co]bleomycin (A, A, B)	Hall et al., 1977
MCA-induced cerebral sarcoma	Yale-Swiss	[^{99m}Tc]pertechnetate -Sn citrate -Sn DTPA -Fe ascorbic acid -Fe ascorbic acid-DTPA	Haynie et al., 1977
MCA-brain sarcoma	Yale-Swiss	[^{111}In]chloride -serum proteins -transferrin	Konikowski et al., 1977
CT-1 fibrosarcoma	mice	[^{99m}Tc]cyclophosphamide	Ercan et al., 1977

TABLE 4-5. (Continued)

Tumor	*Host*	*Radiotracer*	*Reference*
L1210 leukemia	mice	[5-^{18}F]fluorouridylate	Friedkin et al., 1978
Ependymoblastoma	mice	[5-^{18}F]fluorouridylate	Friedkin et al., 1978
EMT-6 sarcoma	BALB/c	[^{67}Ga]citrate	Larson et al., 1978
Sarcoma-180	mice	[^{67}Ga]citrate [^{111}In]chloride [^{67}Ga & ^{111}In]oxine complexes (bleomycin, trasylol, streptokinase and plasminogen)	Rayudu et al., 1978
Human neuroblastoma	nude mice	[^{55}Co/^{57}Co]bleomycin [^{57}Co]chloride [^{67}Ga]citrate [^{131}I]nor-cholesterol	Yeh and Helson, 1978
Human malignant Schwannoma	nude mice	as above	Yeh and Helson, 1978
Human testicular adenocarcinoma	nude mice	as above	Yeh and Helson, 1978
Human non-Hodgkins lymphoma	nude mice	as above	Yeh and Helson, 1978
RILQ lymphoma	CBA	[^{67}Ga]citrate [^{59}Fe]citrate	Sephton et al., 1978 Sephton et al., 1978
WEH17 lymphoma	CBA	as above	Sephton et al., 1978
HPC 108 myeloma	BALB/c	as above	Sephton et al., 1978
P3K myeloma	BALB/c	as above	Sephton et al., 1978
Virus-induced mammary adenocarcinoma	XVII/Bln	[^{167}Tm]citrate [^{67}Ga]citrate	Bayer et al., 1978
Cloudman melanoma S-91	DBA2J	[^{67}Ga]citrate	Hoffer et al., 1979
EMT-6 sarcoma	BALB/c	[^{67}Ga]citrate [^{59}Fe]citrate	Larson et al., 1979a

TABLE 4-5. (Continued)

Tumor	*Host*	*Radiotracer*	*Reference*
EMT-6 sarcoma	mice	[^{67}Ga]citrate [^{97}Ru]ionic -oxine-7-carboxylic acid acetate -*cis*-dichlorotetraamine Ru(III) chloride -other Ru(II, III, & IV) compounds	Srivastava et al., 1979a
L1210 lymphoid	CDF1	[5-^{18}F]U	Shani et al., 1978
P1798 lymphocytic	BALB/c	[5-^{18}F]U	Shani et al., 1978
Ca 755 mammary	C57/BL/6	[5-^{18}F]U	Shani et al., 1978
Ehrlich ascites	Swiss	[5-^{18}F]U	Shani et al., 1978
Sarcoma 180	Swiss	[5-^{18}F]U	Shani et al., 1978
Ridgeway osteogenic sarcoma	AKR	[5-^{18}F]U	Shani et al., 1978
Ehrlich ascites	Swiss	[^{3}H]5FU	Abrams et al., 1979
Lewis Lung	BDF_1	[^{3}H]5FU	Abrams et al., 1979
Ehrlich ascites	Swiss	[^{3}H]5UdR	Abrams et al., 1979
Lewis Lung	BDF_1	[^{3}H]5FUdR	Abrams et al., 1979
Lewis Lung	BDF_1	[2^{1}-^{82}Br]UdR	Lee et al., 1979a
Tumor	C3HBA	[^{125}I]DNA	Hapke and Ithakissios, 1979
Non-Hodgkins lymphoma	nude mice	as above	Yeh et al., 1979
DMBA squamous cell	NMRI	[^{57}Co]bleomycin	Pertynski et al., 1979
Mammary carcinoma	mice	[*I]DES-diphosphate	Mende et al., 1979
Mammary carcinoma	mice	[^{75}Se]purine nucleosides (several)	Deckart et al., 1979
Ehrlich ascites	mice	as above	Deckart et al., 1979

TABLE 4-5. (Continued)

Tumor	*Host*	*Radiotracer*	*Reference*
Sarcomas (3)	mice	as above	Deckart et al., 1979
Ehrlich ascites	Swiss	[^{58}Co]bleomycin -A bleomycin -citrate	Rembelska et al., 1979
NK/Ly lymphoma	Swiss	as above	Rembelska et al., 1979
Harding-Passey melanotic melanoma	NMRI	[^{131}I]tyrosine derivatives (21 compounds)	Kloss and Levine, 1979
Malignant melanoma	mice	[^{201}Tl]chloride	Hudson et al., 1979
Rhodesia tumor	mice	[^{201}Tl]chloride	Hudson et al., 1979
EMT-6 sarcoma	BALB/c	[^{67}Ga]transferrin	Larson et al., 1979b
SAD_2 fibrosarcoma	mice	[^{64}Cu]3-ethoxy-2-oxobutyraldehydebis(thiosemicarbazone)	Pastakia et al., 1980
MTG-B mammary	mice	as above	Pastakia et al., 1980
MCA-induced C3H/HeJ sarcoma	DBA/2J	[^{131}I]alloantibody	Bushman et al., 1980
P1798 lymphosarcoma	CD2F	[^{67}Ga]citrate [^{68}Ga]citrate [^{46}Sc]citrate	Hayes et al., 1980
SaD_2 fibrosarcoma	DBA/2	as above	Hayes et al., 1980
P3/X63-Ag8 myeloma	BALB/c	[^{131}I]hybridoma monoclonal antibody	Levine et al., 1980a
MH-15 teratocarcinoma	BALB/c	as above	Levine et al., 1980a
EMT-6 sarcoma	BALB/c	[^{67}Ga]transferrin	Larson et al., 1980a
EMT-6 sarcoma	BALB/c	[^{67}Ga]citrate	Oster et al., 1980
Spontaneous adenocarcinoma	mice	[^{67}Ga]citrate	Oster et al., 1980
MCA-induced fibrosarcoma	C57B	[^{67}Ga]citrate	Hardoff et al., 1980
EMT-6 sarcoma	mice	[^{3}H]thymidine -uridine [^{14}C]2-deoxyglucose [^{67}Ga]citrate	Larson et al., 1980b

TABLE 4-5. (Continued)

Tumor	*Host*	*Radiotracer*	*Reference*
Sarcoma 180	NMRI	[^{131}I]DNase	Reske et al., 1980
KHJJ	BALB/c	[^{111}In]BLEDTA -benzyl BLEDTA -chloride [^{57}Co]bleomycin A-2 -bleomycin A-2DM [^{67}Ga]citrate	Baumert et al., 1980
Rideway osteogenic sarcoma	AKD2-F1	[^{57}Co]bleomycin -pepleomycin -bleomycin acid -tallysomycin -phleomycin	Hall et al., 1980
MH-15 tetracarcinoma	BALB/c	[^{131}I]tumor specific monoclonal antibodies	Levine et al., 1980b
EMT-6 sarcoma	mice	[^{97}Ru]transferrin	Srivastava et al., 1980
Murine leukemia	HRS/J	[2-^{18}F]DG	Som et al., 1980
Spontaneous adenocarcinoma	BNL	[2-^{18}F]DG	Som et al., 1980
Transplantable adenocarcinoma	BNL	[2-^{18}F]DG	Som et al., 1980
L1210 leukemia	CDF_1	[2-^{18}F]DG	Som et al., 1980
P-1798 lymphosarcoma	Swiss	[^{125}I]DNA	Ithakissios, 1980
CA-755	Swiss	[^{125}I]DNA	Ithakissios, 1980
B-16 melanoma	C57BL/6J	[^{125}I]DNA	Ithakissios, 1980
Ehrlich carcinoma	DDN	[^{125}I]DNA	Ithakissios, 1980
Sarcoma	mice	[^{99}Tc^{m}]5′-guanylyl-β,γ-imino DP -5′-adenylyl-β,γ-imino DP -α,β-methyleneadenosine-5′-T -β,γ-methyleneadenosine-5′-TP	Hosain et al., 1980
Neuroblastoma	mice	as above	Hosain et al., 1980

TABLE 4-5. (Continued)

Tumor	*Host*	*Radiotracer*	*Reference*
SR-DDD-BR Rous-induced sarcoma	SS-DDD	[^{57}Co]bleomycin	Kakinuma et al., 1980
ADJ/PC6 tumor	BALB/c	[^{67}Ga]citrate	Hammersley et al., 1980
S-180	BALB/c	[^{67}Ga]citrate	Hammersley et al., 1980
B-16 melanoma	BALB/c	[^{67}Ga]citrate	Hammersley et al., 1980
Lewis Lung	BDF_1	[^{153}Sm]transferrin -citrate -chloride	Tse et al., 1980
Sarcoma 37	BALB/c	[^{99m}Tc]adriamycin	Clode et al., 1980
Lewis Lung	BDF_1	[6-^{82}Br]U	Wiebe et al., 1980
MH-15 teratocarcinoma	BALB/c	[^{131}I]monoclonal antibodies -F(ab')2μ fragments	Levine et al., 1981
L1210 lymphocytic	mice	[2-^{14}C]5FU [2-^{14}C]5FU	Russ and Campione, 1981
AKR lymphoma	CD-1	[^{99m}Tc]minimicro aggregated albumin	Gallagher et al., 1981
AKR lymphoma	AKR/J	as above	Gallagher et al., 1981
Ehrlich tumor	mice	[^{99m}Tc]DMSA	Yokoyama et al., 1981
EMT-6	BALB/c	[^{3}H]thymidine	Larson et al., 1981a
EMT-6	BALB/c	[^{3}H]2-deoxyglucose	Larson et al., 1981b
C HBA	C H	[^{67}Ga]citrate	Chilton et al., 1981
Lewis Lung	C Bl	[^{67}Ga]citrate	Chatal et al., 1981
LX-1	human/ Orlnu athymic nude mice	[^{67}Ga]citrate	Chatal et al., 1981
PC6 plasmacytoma	BALB/c	[2-^{14}C]thymidine [5-^{125}I]UdR	Robbins and Taylor, 1981
UVT 15264 sarcoma	XVII/Berlin	[^{131}I]anticlostridium antibodies	Altenbrunn, 1981

TABLE 4-5. (Continued)

Tumor	*Host*	*Radiotracer*	*Reference*
Ehrlich tumor	ddy	[^{111}In]citrate [^{169}Yb]citrate	Ando et al., 1981
Ehrlich ascites tumor	ddy	[U-^{14}C]glucose	Kojima et al., 1981
melanoma	mice	[$^{125/131}$I]4-amino-7-iodo-quinolines (8 compounds)	Goulding et al., 1979
Lewis Lung	BDF_1	[2′-^{82}Br]UdR	Lee et al., 1979b
LAF-1 mastocytoma	mice	[^{125}I]5-iodotryptophan -6-iodotryptophan -4-iodophenylalanine -4-iodoamphetamine	Manger et al., 1981
TGS mammary	mice	[^{67}Ga]citrate	Shukla et al., 1981
EMT-6 sarcoma	BALB/c	[^{103}Ru/^{97}Ru]complexes (7 compounds) [^{67}Ga]citrate [^{99m}Tc]DTPA -MDA	Srivastava et al., 1979b
KHJJ adenosarcoma	BALB/c	[^{111}In]BLEDTA	Goodwin et al., 1979
Brain tumors	Yale-Swiss	[^{51}Cr]RBC's [^{131}I]albumin [^{35}S]sulfate	Glenn et al., 1979
EMT-6 sarcoma	BALB/c	[^{67}Ga]citrate	Larson et al., 1979c
Ehrlich ascites	ICR	[^{3}H]FUdR	Abrams et al., 1978
Sarcoma-180	mice	[5-^{211}At]uracil [5-^{211}At]deoxyuridine	Rossler et al., 1979
Sarcoma-180	mice	[^{181}Hf]citrate [^{99m}Tc*I]serum proteins -HSA	Friedman et al., 1979
EMT-6 sarcoma	BALB/c	[^{3}H]thymidine -uridine [^{14}C]2-deoxyglucose [^{67}Ga]citrate	Larson et al., 1980c
EMT-6 sarcoma	BALB/c	[^{67}Ga]citrate	Larson et al., 1979d
PC6 plasma-cytoma	BALB/c	[^{67}Ga]citrate	Hammersley and Taylor, 1980

TABLE 4-5. (Continued)

Tumor	*Host*	*Radiotracer*	*Reference*
renal carcinoma	BALB/c	[^{67}Ga]citrate	Hammersley and Taylor, 1980
S-180	BALB/c	[^{67}Ga]citrate	Hammersley and Taylor, 1980
Sarcoma PM (spontaneous)	BALB/c	[^{67}Ga]citrate	Hammersley and Taylor, 1980
A15	BALB/c	[^{67}Ga]citrate	Hammersley and Taylor, 1980
P-1798 lymphoma	BALB/c	[^{111}In]bleomycin	Sartiano et al., 1980
S-180 sarcoma	mice	[^{67}Ga]citrate -chloride -transferrin	Hammersley and Zivanovic, 1980
ADJ/PC6 plasmacytoma	mice	as above	Hammersley and Zivanovic, 1980
6C3HED lymphosarcoma	mice	[^{67}Ga]citrate	Anghileri et al., 1976
BW5147 lymphatic leukemia	mice	[^{67}Ga]citrate	Anghileri et al., 1976
Ehrlich ascites	mice	[^{57}Co/^{99m}Tc/^{51}Cr/^{111}In] bleomycin	Zimmerman and Hale, 1976
BW10232 adenocarcinoma	C57BL/6J	[^{57}Co]hematoporphyrin	Anghileri et al., 1976b
BW5147 lymphatic leukemia	AKR/J mice	[^{131}I]UdR -CdR	Anghileri and Heidbreder, 1976
Ehrlich ascites	CFI	[^{121}I]chlorobucil [^{57}Co]bleomycin [^{103}Ru]metalocenes (2)	Wenzel et al., 1979
EL4	C57B16	as above	Wenzel et al., 1979
Spontaneous ovarian	C3HeB/FeJ	[^{211}At]tellurium colloid	Bloomer et al., 1981

TABLE 4-6. Uptake of Radiotracer by Rat Tumors, 1976–1981

Tumor	*Host*	*Radiotracer*	*Reference*
7777 hepatoma	Buffalo	[^{131}I]tetracycline [^{67}Ga]citrate [^{99}Tc^{m}]pertechnetate -glucoheptonate -tetracycline -phosphate -pyrophosphate	Chauncey et al., 1976
13762 mammary adenocarcinoma	Fischer-344	[^{57}Co]bleomycin [^{125}I]bleomycin	Eckelman et al., 1976
liver tumor	rat	[^{13}N]BCNU	Pettit et al., 1976
colon tumor	rat	[^{13}N]BCNU	Pettit et al., 1976
RC6 tumor	rat	[^{99m}Tc]3-amino-3-carboxypropyl phosphonic acid -2-aminoethylphosphonate -aminoethylphosphonate -2-aminophosphonobutyric acid 1-4-dihydroxybutyldiene-1,1-4-4-teraphosphonic acid	Castronovo et al., 1976
tumor induced by N-2-fluorenyldiacetamine and radiation source	Charles River CD	X-rays ^{113m}In ^{198}Au	Witcofski and Pizzarello, 1976
RFT tumor	rat	[^{11}C]1-aminocyclopentane-carboxylic acid	Hayes et al., 1976
R-3259 giant cell sarcoma	rat	as above	Hayes et al., 1976
Morris hepatoma 5123c	rat	as above	Hayes et al., 1976
Morris hepatoma 7794	rat	as above	Hayes et al., 1976
Morris hepatoma 7787	rat	as above	Hayes et al., 1976
Morris hepatoma 7777	rat	as above	Hayes et al., 1976
Novikov hepatoma	rat	as above	Hayes et al., 1976
H-35 Reuber hepatoma	rat	as above	Hayes et al., 1976

TABLE 4-6. (Continued)

Tumor	*Host*	*Radiotracer*	*Reference*
Diethylnitrosamine-induced hepatic tumor	Sprague-Dawley	[^{64}Cu]myambutol [^{61}Cu]myambutol	Schumacher et al., 1976
Walker 256 carcinosarcoma	Sprague-Dawley	[^{67}Ga]citrate	Bradley et al., 1976
Glioblastoma	rat	[^{67}Ga]citrate	Davis et al., 1977
Morris hepatoma 7316A	Buffalo	[^{67}Ga]citrate [^{111}In]chloride	Takeda et al., 1976
Morris hepatoma 7777	Buffalo	[^{46}Sc]chloride [^{46}Mn]chloride [^{65}Zn]chloride [^{111}In]chloride [^{195}Au]chloride [^{67}Ga]citrate	Hagan et al., 1977
Walker 256 carcinosarcoma	Sprague-Dawley	[^{67}Ga]citrate	Bradley et al., 1976
Walker 256 carcinosarcoma	rat	[^{3}H]adriamycin	Chang et al., 1978
Walker 256 carcinosarcoma	Sprague-Dawley	[5^{18}F]U	Shani et al., 1978
Novikoff hepatoma	Sprague-Dawley	[5^{18}F]U	Shani et al., 1978
Murphy-Sturm lymphosarcoma	Sprague-Dawley	[5^{18}F]U	Shani et al., 1978
R-3230AC mammary adenosarcoma	Fischer-344	[5^{18}F]U	Shani et al., 1978
13672 mammary adenosarcoma	Fischer-344	[5^{18}F]U	Shani et al., 1978
Dunning cytoxan leukemia	Fischer-344	[5^{18}F]U	Shani et al., 1978
Dunning-O leukemia	Fischer-344	[5^{18}F]U	Shani et al., 1978
Morris hepatoma 7800	Buffalo	[5^{18}F]U	Shani et al., 1978
Morris hepatoma 7777	Buffalo	[^{115m}Cd]chloride [^{125}I]HSA [^{57}Co]bleomycin [^{125}I]HSA [^{75}Se]selenite -selenomethionine	Chauncey et al., 1978
Walker 256 sarcoma	Sprague-Dawley	[^{67}Ga]citrate	Bradley et al., 1979

TABLE 4-6. (Continued)

Tumor	*Host*	*Radiotracer*	*Reference*
Morris hepatoma 5123	Sprague-Dawley	[^{125}I]DNA	Hapke and Ithakissios, 1979
Morris hepatoma 7777	Buffalo	[^{14}C]DL-lactic acid	Chandra and Freed, 1979
Swarm chondrosarcoma	rat	[^{14}C]hexamethonium [^{125}I]hexamethonium	Korn et al., 1979
Mammary carcinoma	rat	*I-DES diphosphate	Mende et al., 1979
Walker 256	rat	[^{99m}Tc]liposomes	Jeyasingh et al., 1979
Walker 256	rat	[2′-^{123}I]UdR	Abrams et al., 1979c
Walker 256	rat	[^{125}I]UdR	Grossman et al., 1979
Yoshida sarcoma	rat	[^{57}Co]bleomycin	Raban et al., 1979
Morris Hepatoma 3924A	ACI	[^{67}Ga]6-mercaptopurine -chloro-6-mercapto-purine -citrate -chloride	Guarino et al., 1979a
Rhabdomyosarcoma	Wag/Ry	[^{57}Co]bleomycin [^{125}I]bleomycin [^{57}Co/^{125}I]bleomycin	Vos et al., 1979
CSE fibrosarcoma	rat	[^{64}Cu]3-ethoxy-2-oxobutyraldehyde bis-(thiosemicarbazone)	Pastakia et al., 1980
Sqaumous cell carcinoma	Fischer	as above	Pastakia et al., 1980
Morris hepatoma 5123C	Buffalo	[^{67}Ga]citrate [^{68}Ga]citrate [^{46}Sc]citrate	Hayes et al., 1980
Morris hepatoma 7777	Buffalo	as above	Hayes et al., 1980
RFT	Fischer 344	as above	Hayes et al., 1980
DMBA-induced mammary	Sprague-Dawley	[^{3}H]*o*-fluorohexestrol [^{3}H]1-fluorohexestrol -1-bromohexestrol -1-iodohexestrol	Katzenellenbogen et al., 1980

TABLE 4-6. (Continued)

Tumor	*Host*	*Radiotracer*	*Reference*
13752 mammary adenocarcinoma	Fischer	[16α-^{125}I]iodoestradiol-17β	Shaughnessey et al., 1980
1372 mammary adenocarcinoma	Fischer	[^{203}Hg]estradiol	Shani et al., 1980
Spontaneous mammary tumors	Sprague-Dawley	[^{203}Hg]estradiol	Shani et al., 1980
13762 mammary adenocarcinoma	Fischer 344/CRBL	[^{99m}Tc]tetrasulfophthalocyanine	Rosseau et al., 1980
RT-1 glioma	rat	[2-^{18}F]DG [2-^{18}F]DG	Goodman et al., 1980
RT-24 glioma	rat	as above	Goodman et al., 1980
RIB5-fibrosarcoma	Johns	[^{125}I]DNA	Ithakissios, 1980
Mammary tumors	Fischer	[^{125}I]DNA	Ithakissios, 1980
Walker W-256 carcinosarcoma	Fischer 344	[^{125}I]DNA	Ithakissios, 1980
Mammary carcinoma	rat	[^{99}Tc^{m}]5′-guanylyl-β,γ-imino DP -5′-adenylyl-β,γ-imino DP -α,β-methyleneadenosine 5′-TP -β,γ-methyleneadenosine 5′-TP	Hosain et al., 1980
Yoshida sarcoma	Donryu	[^{68}Ga]citrate	Muranaka et al., 1980
Renal sarcoma	rat	[^{67}Ga]citrate	Hammersley et al., 1980
Fibrosarcoma	rat	[^{67}Ga]citrate	Hammersley et al., 1980
Dunning 3327H	Copenhagen	[^{153}Sm]transferrin -citrate -chloride	Tse et al., 1980
Walker 256	rat	[6-^{123}I]U	Wiebe et al., 1980
DMBA-induced mammary	Sprague-Dawley	[16α-^{77}Br]bromoestradiol-17β	Katzenellenbogen et al., 1981a
Morris hepatoma 5123C	Buffalo	[^{67}Ga]citrate	Hayes et al., 1981

TABLE 4-6. (Continued)

Tumor	*Host*	*Radiotracer*	*Reference*
13762 mammary adenocarcinoma	Fischer	[^{125}I]iodoestradiol-17β	Gatley et al., 1981
Spontaneous mammary	Sprague-Dawley	as above	Gatley et al., 1981
NMU-induced	Sprague-Dawley	as above	Gatley et al., 1981
DMBA-induced mammary	rat	[16α^{77}Br]bromoestradiol-17β	McElvaney et al., 1981
Mammary tumor	rat	[^{77}Br]estrogen	Katzenellenbogen, 1981b
TCT-4904 bladder transitional cell carcinoma	Fischer	[^{57}Co]porphyrins	Fawwaz et al., 1981
Morris heptaoma 7777	Buffalo	[^{3}H]thymidine	Larson et al., 1981a
Morris hepatoma 7777	Buffalo	[^{3}H]2-deoxyglucose	Larson et al., 1981b
Yoshida sarcoma	Donryu	[^{111}In]citrate [^{169}Yb]citrate	Ando et al., 1981
Hepatoma AH109A	Donryu	as above	Ando et al., 1981
Dunning R3327H	Copenhagen	[^{11}C]α-aminoisobutyric acid	Dunzendorfer et al., 1981
Dunning R3327G	Copenhagen	as above	Dunzendorfer et al., 1981
Walker 256	Wistar	[2′-^{123}I]UdR	Abrams et al., 1981
Walker 256	rat	[^{67}Ga]transferrin	Vallabhajosula et al., 1979
Hepatoma	rat	[^{67}Ga]citrate [^{54}Mn]chloride	Halpern and Hagan, 1979a,b
Hepatoma	rat	[^{67}Ga]citrate	Hayes et al., 1979
Squamous cell carcinoma	rat	[^{67}Ga]citrate [^{97}Ru]ionic -oxine-7-carboxylic acid acetate -*cis*-dichlorotetraamine Ru III chloride -other Ru (II,III,IV) compounds	Srivastava et al., 1979a

TABLE 4-6. (Continued)

Tumor	*Host*	*Radiotracer*	*Reference*
Morris hepatoma 5123	Buffalo	1-aminocyclobutane[^{11}C/^{14}C]carboxylic acid	Washburn et al., 1979
Morris hepatoma 7777	Buffalo	as above	Washburn et al., 1979
Reuber H-35 hepatoma	Buffalo	as above	Washburn et al., 1979
Morris hepatoma 39244A	ACI	[^{67}Ga]6-mercaptopurine-citrate -chloride [^{67}Ga]citrate -chloride	Guarino et al., 1979b
Morris Hepatoma	ACI	[^{67}Ga]chloride -citrate	Shukla, 1979
DMBA-induced mammary adenocarcinoma	Sprague-Dawley	[^{77}Br]16α-bromoestradiol-17β	Katzenellenbogen, 1981c
Spontaneous mammary	Sprague-Dawley	[^{203}Hg]estradiol	Shani et al., 1981
13762 mammary adenocarcinoma	Fischer	[^{203}Hg]estradiol	Shani et al., 1981
Walker 256	rat	[^{67}Ga]citrate	Vallabhajosula et al., 1981
Murphystum lymphosarcoma	rat	[^{67}Ga]citrate	Vallabhajosula et al., 1981
Morris hepatoma 3924A	rat	[^{67}Ga]citrate	Shukla et al., 1981
Walker 256	rat	[^{131}I]6-iodo-deoxygalactose	Flanagan et al., 1979
Squamous-cell carcinoma	Fischer CDF	[^{103}Ru/^{97}Ru]complexes (7 compounds) [^{67}Ga]citrate [^{99}Tc^{m}]DTPA -MDP	Srivastava et al., 1979b
Hepatoma	Buffalo	[^{123}I]iodobenzamides (5 compounds)	Baldwin et al., 1981
Yoshida sarcoma	Sprague-Dawley	[^{131}I]albumin macroaggregates	Klinger et al., 1976
Hepatoma	Sprague-Dawley	[^{131}I]albumin macroaggregates	Klinger et al., 1976

TABLE 4-7. Uptake of Radiotracer by Canine Tumors, 1976–1981

Tumor	*Radiotracer*	*Reference*
Spontaneous neoplasm (5)	[*I]fibrinogen	Wortman et a., 1976
Spontaneous neoplasms	[^{13}N]ammonia -L-glutamine	Gelbard et al., 1976
TVT	[^{67}Ga]citrate -transferrin [^{68}Ga]citrate [^{125}I]transferrin	Noujaim et al., 1978
Spontaneous seminoma	^{18}FDG [^{18}FDG]6-phosphate	Som et al., 1978
Spontaneous adreno-cortical tumors	[^{131}I]19-cholesterol	van den Brom, 1979
TVT	[^{67}Ga]transferrin [^{68}Ga-^{125}I]transferrin [^{68}Ga]citrate	Wong et al., 1980
TVT	[^{67}Ga]citrate	Hammersley et al., 1980
Spontaneous lymphoma	[^{3}H]thymidine	Larson et al., 1981a
Spontaneous lymphoma	[^{3}H]thymidine	Larson et al., 1981a
Soft-tissue sarcoma	[^{3}H]thymidine	Larson et al., 1981a
Thyroid carcinoma	[^{3}H]thymidine	Larson et al., 1981a
Spontaneous lymphoma	[^{3}H]2-deoxyglucose	Larson et al., 1981b
Spontaneous lymphoma	[^{3}H]2-deoxyglucose	Larson et al., 1981b
Soft-tissue sarcoma	[^{3}H]2-deoxyglucose	Larson et al., 1981b
Thyroid carcinoma	[^{3}H]2-deoxyglucose	Larson et al., 1981b
TVT	[^{67}Ga-^{125}I]transferrin [^{67}Ga]transferrin [^{68}Ga]citrate	Terner et al., 1979
Osteogenic sarcomas 4	[^{13}N]glutamic acid	McDonald et al., 1976
Epidermoid carcinoma/ metastases	[^{11}C]methylputrescine	Miller et al., 1978
Hepatoma	[^{11}C]methylputrescine	Miller et al., 1978
Mammary adenocarcinoma/ metastases	[^{11}C]methylputrescine	Miller et al., 1978
Fibroma	[^{11}C]methylputrescine	Miller et al., 1978
Malignant mammary tumor	[^{11}C]methylputrescine	Miller et al., 1978

TABLE 4-8. Uptake of Radiotracer by Hamster Tumors, 1976–1981

Tumor	*Host*	*Radiotracer*	*Reference*
Kirkman 46710 pancreatic duct adenocarcinoma	Syrian Golden	[^{99m}Tc]1-aminocyclopentane	Heindel et al., 1976
Kirkman 2309V islet cell adenocarcinoma	Syrian Golden	as above	Heindel et al., 1976
Pancreatic tumor	hamster	[^{35}S]thio-D-glucose	Markoe et al., 1977
Amelanotic melanoma Fortner III	golden	[^{131}I]UdR	Volm et al., 1977
Greene melanoma	hamster	[^{67}Ga] [^{97}Ru]ionic -oxine-7-carboxylic acid acetate -*cis* dichlorotetraamine Ru (III) chloride -other Ru (II,III,IV) compounds	Srivastava et al., 1979a
46710 pancreatic duct adenocarcinoma	hamster	1-aminocyclobutane [^{11}C/^{14}C]-carboxylic acid	Washburn et al., 1979a
2309V pancreatic islet adenocarcinoma	hamster	as above	Washburn et al., 1979a
46710 pancreatic duct adenocarcinoma	Syrian Golden	DL[carboxyl ^{11}C/^{14}C]-tryptophan	
2309V pancreatic islet adenocarcinoma	Syrian Golden	as above	Washburn et al., 1979b
Pancreatic adenocarcinoma Pan No. 1	Syrian Golden	as above	Washburn et al., 1979b
46710 pancreatic duct adenocarcinoma	Syrian Golden	[^{35}S]5-thioglucose	Markoe et al., 1979
2309V pancreatic islet adenocarcinoma	Syrian Golden	[^{35}S]5-thioglucose	Markoe et al., 1979
Pancreatic adenocarcinoma Pan No. 1	Syrian Golden	[^{35}S]5-thioglucose	Markoe et al., 1979
Human GW-39 colonic tumor	Syrian Golden	[^{125}I]DNA	Hapke and Ithakissoios, 1979b
Mohr melanoma	hamster	[^{131}I]tyrosine derivatives (21 compounds)	Kloss and Levine, 1979
Fortner melanoma	hamster	as above	Kloss and Levine, 1979

TABLE 4-8. (Continued)

Tumor	*Host*	*Radiotracer*	*Reference*
Cheek-pouch carcinoma	hamster	[^{99m}Tc]5′-guanylyl-β,γ-imino DP -5′-adenylyl-β,γ-imino DP -α,β-methylene-diamine 5′-TP -β,γ-methylene-diamine 5′-TP	Hosain et al., 1980
Greene melanoma	Syrian Golden	[^{14}C]L-tyrosine -DOPA [DL-^{11}C]DOPA	van Langevelde et al., 1981
Greene melanoma	Syrian Golden	[^{103}Ru/^{97}Ru] complexes (7 compounds) [^{67}Ga]citrate [^{99m}Tc]DTPA -MDP	Srivastava et al., 1979b
Greene melanoma	Syrian Golden	Na[^{32}P]phosphate [^{111}In]bleomycin -chloride [^{67}Ga]citrate [^{203}Pb]TRIS [^{197}Hg]chlormerodrin [^{201}Tl]chloride [^{123}I]quinoline [^{57}Co]bleomycin	Packer et al., 1977
Greene melanoma	Syrian Golden	[^{123}I]indocyanine green	Ansari et al., 1976
Green melanoma	Syrian Golden	[^{203}Pb]TRIS [^{57}Co]bleomycin [^{201}Tl]chloride	Packer et al., 1976
Melanotic melanoma	Syrian Golden	[^{131}I]4-(3-DMAPA)-7-iodoquinoline [^{131}I]3-iodo-α-methyl-tyrosine	Bubeck et al., 1981

TABLE 4-9. Uptake of Radiotracer by Rabbit Tumors, 1976–1981

Tumor	*Host*	*Radiotracer*	*Reference*
Brown-Pearce carcinoma	rabbit	[^{131}I]UdR	Volm et al., 1977
VX2	New Zealand	[^{11}C]thymidylate -thymidine	Crawford et al., 1978
VX-2 epidermoid	rabbit	[^{201}Tl]chloride [^{67}Ga]citrate [^{42}K]chloride	Ito et al., 1978
X-ray-induced osteosarcoma	New Zealand	[^{99m}Tc]pyrophosphate	King et al., 1980
Spontaneous adenocarcinoma	rabbit	[^{67}Ga]citrate	Oster et al., 1980
VX-2 epidermoid carcinoma	rabbit	[^{99m}Tc]sulfur colloid	Otsuka et al., 1981

An analysis of the data in Tables 4-5 through 4-10 to determine the suitability of particular animal tumor models for screening specific radiotracers is unfortunately not possible. Authors tend not to rationalize the basis on which the tumor model was selected, nor do they interpret their data in terms of specific tumor characteristics. Greater contributions to progress toward the development of diagnostic radiotracers for oncology are dependent in no small part on a rational basis for selection of the animal tumor

TABLE 4-10. Uptake of Radiotracer in Unnamed Hosts, 1976–1981

Tumor	*Radiotracer*	*Reference*
Spontaneous	[*I]fibrinogen	DeNardo et al., 1976b
Neurogenic	[^{13}N]BCNU	Pettit et al., 1976
Nonneurogenic	[^{13}N]BCNU	Pettit et al., 1976
Walker 256	[^{14}C]ornithine	Buffkin and Webber, 1978
MCA-induced tumor	[^{14}C]ornithine	Buffkin and Webber, 1978
Spontaneous adenocarcinoma	[^{18}F]DG [^{18}FDG]6-phosphate	Som et al., 1979
L-1210	as above	Som et al., 1979
Spontaneous leukemia	as above	Som et al., 1979
Ependyomona	as above	Som et al., 1979
Pituitary tumor	as above	Som et al., 1979
Fibroamelanotic melanoma	as above	Som et al., 1979
Tumor	[^{131}I]anticlostridium antibodies	Altenbrun et al., 1979

model used in the testing program. Fortunately, some progress in this area is evident, especially in those investigations that deal with receptor-oriented or immunologically active radiopharmaceuticals.

ACKNOWLEDGMENTS

This review is based on lectures presented to the mid-winter (Phoenix, January, 1982) and annual (Miami, June, 1982) meetings of the Radiopharmaceutical Science Council of the Society of Nuclear Medicine. I wish to express my appreciation to the Council, in particular to Dr. W. Eckelman and Dr. R. M. Lambrecht for the invitation to participate. The contribution of my research colleagues and staff, and the secretarial assistance of Mrs. Gladys McIntyre are gratefully acknowledged.

REFERENCES

Abrams, D. N., Knaus, E. E., McQuarrie, S. A., and Wiebe, L. I. (1978). [^{18}F]5-fluoro-2′-deoxyuridine as a radiopharmaceutical for diagnostic oncology. *The Chemistry of Radiopharmaceuticals,* N. D. Heindel, H. D. Burns, T. Honda, and L. W. Brady, eds. New York: Masson, pp. 205–213.

Abrams, D. N., Knaus, E. E., and Wiebe, L. I. (1979a). Tumor uptake of radiolabeled pyrimidine bases and pyrimidine nucleosides in animal models I. 6-[^{3}H]-5-fluorouracil. *Int. J. Nucl. Med. Biol.* **6**:97–101.

Abrams, D. N., Knaus, E. E., and Wiebe, L. I. (1979b). Tumor uptake of radiolabeled pyrimidine bases and pyrimidine nucleosides in animal models II. 6-[^{3}H]-5-fluoro-2′-deoxyuridine. *Int. J. Nucl. Med. Biol.* **6**:103–107.

Abrams, D. N., Knaus, E. E., Wiebe, L. I., Helus, F., and Maier-Borst, W. (1979c). Tissue uptake and excretion of 2′-[^{123}I]-iodo-2′-deoxyuridine in animal tumor models. *Eur. J. Nucl. Med.* **4**:134.

Abrams, D. N., Knaus, E. E., Wiebe, L. I., Helus, F., and Maier-Borst, W. (1981). Tumor uptake of radiolabelled pyrimidine bases and pyrimidine nucleosides in animal models. IV. 2′-^{123}I-iodo-2′-deoxyuridine. *Int. J. Appl. Radiat. Isot.* **32**:105–108.

Altenbrun, H. J., Mehnert, A. H., Vogt, R., Schmidt, E., and Steenbeck, L. (1979). Positive tumourscintigraphy in animals with [^{131}I]-anticlostridium antibodies. *Eur. J. Nucl. Med.* **4**:134.

Altenbrunn, H. J., Mehnert, W. H., Vogt, R., Schmidt, H. E., and Steenbeck, L. (1981). Scintigraphic tumor localization in mice with radioiodinated anti-clostridium antibodies. *Int. J. Nucl. Med. Biol.* **8**:90–93.

Altman, P. L., and Katz, D. D. (1979). *Inbred and Genetically Defined Strains of Laboratory Animals,* Parts 1 and 2. Bethseda: Federation of the American Societies of Experimental Biology.

Ambrosie, E. J., and Roe, F. J. C. (1975). *Biology of Cancer,* 2nd ed. Westergate, England: Ellis Horwood Ltd.

Ando, A., Ando, I., Takeshita, M., Hiraki, T., and Hisada, K. (1981). Subcellular distribution of ^{111}In and ^{169}Yb in tumor and liver. *Eur. J. Nucl. Med.* **6**:221–226.

Anghileri, L. J., and Heidbreder M. (1976). [^{131}I]deoxyuridine and [^{131}I]deoxycytidine accumulation by tumors. *Nuklearmedizin* **XV**:254–255.

Anghileri, L. J., Heidbreder, M., and Mathes, R. (1976a). Effects of enzymatic treatment on radioactivity accumulation by tumors. *Nuklearmedizin* **XV**:305–307.

Anghileri, L. J., Heidbreder, M., and Mathes, R. (1976b). [^{57}Co]hematoporphrin accumulation by experimental tumors. *Nuklearmedizin* **XV**:183–184.

Ansari, A., Lambrecht, R. M., Packer, S., Atkins, H. L., Redvanly, C. S., and Wolf, A. P. (1976). Note on the distribution of iodine-123 labeled indocyanine green in the eye. XVIII. *Invest. Ophth.* **14**:780–782.

Baldwin, R. M., Lin, T. H., and Winchell, H. S. (1981). Structure-activity relationships of [^{123}I] labelled *o*-iodobenzamide derivatives. *J. Rad. Chem.* **65**:269–277.

Baumert, J. E., Diamanti, C. I., Goodwin, D. A., DeRiemer, L. H., and Meares, C. F. (1980). Comparison of tumor and abscess uptake of [^{111}In]bleomycin analogues with [^{57}Co]bleomycins in BALB/c mice. *J. Nucl. Med.* **21**:36.

Beyer, G.-J., Franke, W.-G., Hennig, K., Johannsen, B. A., Khalkin, V. A., Kretzschmar, M., Lebedev, N. A., Münze, R., Nougorodov, A. F., and Thieme, K. (1978). Comparative kinetic studies of simultaneously injected [^{157}Tm]- and [^{67}Ga]-citrate in normal and tumor bearing mice. *Int. J. Appl. Rad. Isot.* **29**:673–681.

Bloomer, W. D., McLaughlin, W. H., Neirinckx, R. D., Adelstein, S. J., Gordon, P. R., Ruth, T. J., and Wolf, A. P. (1981). Astatine-211-Tellurium Radiocolloid Cures Experimental Malignant Ascites. *Science* **211**:340–341.

Bradley, W. P., Alderson, P. O., Eckelman, W. C., and Weiss, J. F. (1977). [^{67}Ga] tumor uptake in rats after whole body irradiation. *J. Nucl. Med.* **18**:602.

Bradley, W. P., Alderson, P. O., Eckelman, W. C., Hamilton, R. G., and Weiss, J. F. (1978). Decreased tumor uptake of gallium-67 in animals after whole-body irradiation. *J. Nucl. Med.* **19**:204–209.

Bradley, W. P., Alderson, P. O., and Weiss, J. F. (1979). Effect of iron deficiency on the biodistribution and tumor uptake of [^{67}Ga] citrate in animals: Concise communication. *J. Nucl. Med.* **20**:243–247.

Bubeck, B., Eisenhut, M., Heimke, U., and zum Winkel, K. (1981). Melanoma affine radiopharmaceuticals. *Eur. J. Nucl. Med.* **6**:227–233.

Buchsbaum, D. J., Loken, M. K., Johnson, E. A., Walker, P. C., and Ponto, R. A. (1980). Localization of radiolabeled mouse alloantibody in a sarcoma induced by 3-methyl-cholenthrene. *J. Nucl. Med.* **21**:77–80.

Buffkin, D. C., and Webber, M. M. (1978). [^{14}C] ornithine as a marker of malignancy. *J. Nucl. Med.* **19**:731.

Castronovo, F. P., Potsaid, M. S., Kopiwoda, S., and Peterson, M. (1980). Biodistribution of a ^{14}C- and ^{99}Tc-labeled *N*-substituted nitrosourea (CCNU) in an animal tumor model. *J. Rad. Chem.* **60**, 291–300.

Castronovo, F. P., Potsaid, M. S., and Kornblith, P. L. (1976). Technetium-99m phosphonate complexes: Biological distribution in a tumor animal model (RC6). *J. Nucl. Med.* **17**:566.

Ceriani, R. L., Miller, S. W., Thompson, K. E., Peterson, J. A., and Abraham, S. (1976). Labeled cell-specific antibodies in the detection of metastases. *J. Nucl. Med.* **17**:567.

Chandra, R., and Freed, B. R. (1979). Tissue distribution of DL-lactic acid-1[^{14}C] in tumor-bearing rats. *Int. J. Nucl. Med. Biol.* **6**:122–123.

Chang, B., Verma, R. C., Webber, M. M., and O'Connell, T. X. (1978). Localization of intra-arterial vs. intravenous [^{3}H]-adriamycin. *J. Nucl. Med.* **19**:716.

Chatal, J. F., Diez, E., and Guihard, D. (1981). Is the nude mouse bearing a human tumor the best model for study of tumor-localizing radionuclides? *Int. J. Nucl. Med. Biol.* **8**:181–183.

Chauncey, D. M., Halpern, S. E., Hagan, P. L., and Alzaraki, N. P. (1976). Tumor model studies of [^{131}I]-tetracycline and other compounds. *J. Nucl. Med.* **17**:274–281.

Chauncey, D. M., Hagan, P. L., Halpern, S. E., McKegney, M. L., and Bernstein, K. E. (1978). The distribution of cadmium-115m chloride, cobalt-57 bleomycin, iodine-125 human serum albumin, selenium-75 selenite and selenomethionine-75 in a rat hepatoma model. *Eur. J. Nucl. Med.* **3**:243–248.

Chilton, H. M., Witcofski, R. L., Watson, Jr., N. E., and Heise, C. M. (1981). Alteration methotrexate: Concise communication. *J. Nucl. Med.* **22**:1064–1068.

Clode, W. H., Siqueira, M. C., Ferra, M. A., Mendonca, M. A., and Baptista, A. M. (1980). Adriamycin[^{99m}Tc]. A new radioactive agent for the gamma-graphic localization of tumors? A study using as a model BALB/c mice with sarcoma 37 tumors. *Eur. J. Nucl. Med.* **5**:A16.

Comar, D., Cartron, J. C., Maziere, M., and Marazano, C. (1976). Labelling and metabolism of methionine-methyl[^{11}C]. *Eur. J. Nucl. Med.* **1**:11–14.

Crawford, E. J., Christman, D., Atkins, H., Friedkin, M., and Wolf, A. P. (1978). Scintigraphy with positron-emitting compounds I. Carbon-11 labeled thymidine and thymidylate. *Int. J. Nucl. Med. Biol.* **5**:61–69.

Davis, M. A., Taube, R. A., and Carmel, A. D. (1977). Comparison of the biologic distribution of three commercially available [^{67}Ga] citrate preparations in normal and diseased rats. *J. Nucl. Med.* **18**:617.

Deckart, H., Blottner, A., and Spiess, H. (1979). [^{75}Se]-substituted purinnucleosides as potential radiopharmaceuticals for positive tumor scanning. *Eur. J. Nucl. Med.* **4**:134.

DeNardo, G., Krohn, K., DeNardo, S., and Meyers, J. (1976a). Comparison of oncophilic radiopharmaceuticals in tumor bearing rodents. *J. Nucl. Med.* **17**:525.

DeNardo, G., DeNardo, S., Krohn, K., Meyers, J., Huang, S.-M., and Wortman, J. (1976b). Mechanisms of localization of *I-fibrinogen in cancer. *J. Nucl. Med.* **17**:567.

Dunzendorfer, U., Schmall, B., Bigler, R. E., Zanconico, P. B., Conti, P. S., Dahl, J. R., Kleinert, E., Whitmore, W. F. (1981). Synthesis and body distribution of alpha-aminoisobutyric acid-L ^{11}C in normal and prostate cancer-bearing rat after chemotherapy. *Eur. J. Nucl. Med.* **6**:535–538.

Eckelman, W. C., Kubota, H., Siegal, B. A., Komai, T., Rzeszotarski, W. J., and Reba, R. C. (1976). Iodinated bleomycin: An unsatisfactory radiopharmaceutical for tumor localization. *J. Nucl. Med.* **17**:385–388.

Ercan, M. T., Sarizi, T., and Bekdik, C. F. (1977). Labelling and evaluation of [^{99m}Tc]cyclophosphamide for tumor visualization. *Int. J. Appl. Rad. Isot.* **28**:521–528.

Fawwaz, R. A., Wang, T. S. T., and Alderson, P. O. (1981). Evaluation of radioporphyrins as tumor seeking agents in experimental animals. *J. Nucl. Med.* **22**:P50.

Flanagan, R., Lentle, B. C., Wiebe, L. I., Helus, F., and Maier-Borst, W. (1979). Synthesis of 6-radioiodo-6-deoxy-D-galactose, and its tissue distribution and excretion in rats. *Radiopharmaceuticals,* Vol. II. New York: Soc. Nucl. Med., pp. 119–128.

Friedkin, M., Fowler, J., Gallagher, B., MacGregor, R., Wolf, A., Wodinsky, I., and Goldin, A. (1978). 5-[^{18}F]-fluorouridylate as a probe for measuring RNA synthesis and tumor growth rates in vivo. *J. Nucl. Med.* **19:**702.

Friedman, A. M., Sullivan, J. C., Ruby, S. L., Lindenbaum, A., Russell, J. J., Zabransky, B. J., and Rayudu, G. V. S. (1976). Studies of tumor metabolism. I. By use of Mossbauer spectroscopy and autoradiography of [^{153}Sm]. *Int. J. Nucl. Med. Biol.* **3:**37–40.

Friedman, A. M., Rayudu, G. V. S., Sullivan, J. C., Zalutsky, M. R., Ruby, S. L., and Fordam, E. W. (1979). A study of the mechanism of metal ion metabolism by tumors. *J. Lab. Comp. Rad.* **XIII:**250.

Gallagher, B. M., DeLano, M. L., and Camin, L. L. (1981). Radiocolloid uptake in AKR spontaneous lymphoma. *J. Nucl. Med.* **22:**P50–1.

Gatley, S. J., Shaughnessy, W. J., Inhorn, L., and Lieberman, L. M. (1981). Studies with 17 (16α[^{125}I]iodo)-estradiol in estrogen receptor-binding radiopharmaceutical, in rats bearing mammary tumors. *J. Nucl. Med.* **22:**459–464.

Gelbard, A. S., Christie, T. R., Clarke, L. P., and Laughlin, J. S. (1977). Imaging of spontaneous canine tumors with ammonia and L-glutamine labeled with [^{13}N]. *J. Nucl. Med.* **18:**718–723.

Glenn, H. J., Atallah, M. R., Haynie, T. P., Anderson, J. H., and Raulston, G. L. (1979). Compartmental studies of a mouse brain tumor. *Radiopharmaceuticals,* Vol. II. New York: Soc. Nucl. Med., pp. 285–295.

Gold, M. (1981). The cells that would not die. *Science* **81:**29–35.

Goodman, M. M., Elmaleh, D. R., Merk, L., Lade, R., Kearfott, K., Varnum, D., Kopiwoda, S., Brownell, G. L., and Strauss, H. W. (1980). [^{18}F]-2 and 3-fluorodeoxy-D-glucose as potential diagnostic tracers for tumors. *J. Nucl. Med.* **21:**P37.

Goodwin, D. A., Meares, C. F., DeRiemer, L. H., Diamanti, C. I., and Goode, R. L. (1979). [^{111}In] BLEDTA: A conjugate of bleomycin with a bifunctional chelating agent for tumor localization. *Radiopharmaceuticals,* Vol. II. New York: Soc. Nucl. Med., pp. 275–284.

Goulding, R. W., Danpure, H. J., Somaia, S., Osman, S., Gunasekera, S. W., and Eakins, M. N. (1979). Radio-iodine labelled 4-amino-7-iodoquinoline for melanoma detection. *J. Lab. Comp. Rad.* **XVI:**46.

Grossman, V., Kusak, V., and Jarkovska, I. (1979). The [^{125}I]-deoxyuridine as the differential diagnostic of tumor and abscess in rat. *Eur. J. Nucl. Med.* **4:**136.

Guarino, A., Shukla, S. K., Castelli, L., Cipriani, C., and Manni, G. B. (1979a). Uptake of [^{67}Ga]6-mercaptopurines in Morris hepatoma-3924-A. *Eur. J. Nucl. Med.* **4:**211–215.

Guarino, A., Shukla, S. K., Castelli, L., Cipriani, C., and Manni, G. B. (1979b). Preparation and properties of [^{67}Ga]6-mercaptopurine complex and its uptake in Morris hepatoma-3924A. *J. Lab. Comp. Rad.* **XVI:**136.

Hagan, P. L., Chauncey, D. M., Halpern, S. E., and McKegney, M. L. (1977). Comparison of viable and nonviable tumor uptake of Sc-46, Mn-54, Zn-65, [^{111}In] and [^{195}Au] with [^{67}Ga] citrate in a hepatoma model. *Eur. J. Nucl. Med.* **2:**225–230.

Hall, J. N., Chen, J. D., Woolfenden, J. M., Patton, D. D., Liu, R. L., and Salmon, S. E. (1976). Comparative studies of radiolabeled bleomycin and the ionic radiolabeled species in a bleomycin-sensitive tumor model. *J. Nucl. Med.* **17:**567.

Hall, J. N., Woolfenden, J. M., Chen, J. D., and Patton, D. D. (1977). Cobalt-bleomycin A2: A better radiolabeled tumor localizing agent. *J. Nucl. Med.* **18:**617–618.

Hall, J. N., Woolfenden, J. M., Rothman, L. S., and Douros, J. D. (1980). Comparison of tumor-localizing properties of cobalt-57 bleomycin and four analogues: bleomycin acid, phleomycin, pepleomycin and tallysmomycin. *J. Nucl. Med.* **21:**P37.

Halpern, S. E. (1977). Of models and men. *J. Nucl. Med.* **18:**940–942.

Halpern, S. E., and Hagan, P. L. (1979a). Effect of carrier on [^{67}Ga] and [^{54}Mn] in a tumor model. Part 1. *J. Nucl. Med.* **20:**656.

Halpern, S. E., and Hagan, P. L. (1979b). Part 2. Ibid. **20:**656.

Hammersley, P. A. G., and Taylor, D. M. (1980). The effect of the administration of iron on gallium 67 citrate uptake in tumors. *Br. J. Radiol.* **53:**563–571.

Hammersley, P. A. G., and Zivanovic, M. A. (1980). The relationship of [^{67}Ga] uptake to citrate dose, compared with [^{67}Ga]-chloride and [^{67}Ga]transferrin in rodent tissues and tumors. *Nuklearmedizin* **XIX:**25–28.

Hammersley, P. A. G., Taylor, D. M., and Cronshaw, S. (1980). The mechanism of ^{67}Ga uptake in animal and human tumors. *Eur. J. Nucl. Med.* **5:**411–415.

Hayes, R. L., Rafter, J. J., and Byrd, B. L. (1979). Studies on the mechanism of Gallium-67 uptake by tumor, abscess and normal tissue. *J. Nucl. Med.* **20:**672–673.

Hapke, B., and Ithakissios, D. (1979). Radioiodinated nucleic acids for tumor imaging. *Int. J. Nucl. Med. Biol.* **6:**118–121.

Hardoff, R., Front, D., Yoselevski, G., and Ben-Arieh, Y. (1980). Uptake of [^{67}Ga] in irradiated tumors. Is [^{67}Ga] a true indicator of tissue response to irradiation? *J. Nucl. Med.* **21:**P31.

Hayes, R. L., Washburn, L. C., Wieland, B. W., Sun, T. T., Turtle, R. R., and Butler, T. A. (1976). Carboxyl-labeled [^{11}C]-1-aminocyclopentane-carboxylic acid, a potential agent for cancer detection. *J. Nucl. Med.* **17:**748–751.

Hayes, R. L., Byrd, B. L., Rafter, J. J., and Carlton, J. E. (1980). The effect of scandium on the tissue distribution of [^{67}Ga] in normal and tumor-bearing rodents. *J. Nucl. Med.* **21:**361–365.

Hayes, R. L., Rafter, J. J., Byrd, B. L., and Carlton, J. E. (1981). Studies of the *in vivo* entry of [^{67}Ga] into normal and malignant tissues. *J. Nucl. Med.* **22:**325–332.

Haynie, T. P., Konikowski, T., and Glenn, H. J. (1976). Experimental models for the evaluation of radioactive tumor-localizing agents. *Sem. Nucl. Med.* **VI:**347–370.

Haynie, T. P., Konikowski, T., and Glenn, H. J. (1977). Technetium-99m stannous citrate brain-tumor uptake in mice: Concise communication. *J. Nucl. Med.* **18:**915–918.

Heindel, N. D., Risch, V. R., Adams, W. E., Honda, T., and Brady, L. W. (1976). [^{99m}Tc]-1-aminocyclopentane carboxylic acid: Tumor and tissue distribution results on a labeled cytotoxic amino acid. *Int. J. Appl. Radiat. Isot.* **27:**621–625.

Helson, L., Das, S. K., and Hajdu, S. I. (1975). Human neuroblastoma in nude mice. *Cancer Res.* **35:**2594–2599 (1975).

Hewitt, H. B. (1978). The choice of animal tumors for experimental studies of cancer therapy. *Adv. Cancer Res.* **27:**149–200.

Hoffer, P. B., Samuel, A., Bushberg, J. T., and Thakur, M. (1979). Effect of desferoxamine on tissue and tumor retention of gallium-67: Concise communication. *J. Nucl. Med.* **20:**248–251.

Horton, M. B., and Hill, G. J. (1977). *Clinical Oncology*. Philadelphia: W. B. Saunders.

Hosain, P., Zeichner, S. J., Brody, K. R., Hosain, F., and Spencer, R. P. (1980). Studies with technetium-99m labeled nucleotide analogues. *Int. J. Nucl. Med. Biol.* **7**:46–49.

Hudson, F. R., Dewey, D., Galpine, A. R., and Whittingham, A. G. (1979). Tumor uptake of thallium-201 chloride. *Eur. J. Nucl. Med.* **4**:283–284.

Ithakissios, D. S. (1980). Radioiodination of DNA and *in vivo* studies for tumor localization. *Int. J. Nucl. Med. Biol.* **7**:17–24.

Ito, Y., Muranaka, A., Harada, T., Matsudo, A., Yokobayashi, T., and Terashima, H. (1978). Experimental study on tumor affinity of [^{201}Tl]-chloride. *Eur. J. Nucl. Med.* **3**:81–86.

Jeyasingh, K., Jewkes, R. F., Richardson, V. J., Osborne, M. P., and Ryman, B. E. (1979). In-vivo distribution of [^{99m}Tc]-liposomes. *Eur. J. Nucl. Med.* **4**:134.

Kaiser, H. E. (1978). *Neoplasms: Comparative Pathology of Growth in Animals, Plants and Man*. Baltimore: Williams and Wilkins.

Kakinuma, J., Kagiyama, R., and Orii, H. (1980). Chemical properties and tumor affinity of separated isomers of cobalt bleomycin. *Eur. J. Nucl. Med.* **5**:159–163.

Katzenellenbogen, J. A., Carlson, K. E., Heiman, D. F., and Goswami, R. (1980). Receptor-binding radiopharmaceuticals for imaging breast tumors: Estrogen–receptor interactions and selectivity of tissue uptake of halogenated estrogen analogs. *J. Nucl. Med.* **21**:550–558.

Katzenellenbogen, J. A., Senderoff, S. G., McElvany, K. D., O'Brien, H. A., and Welch, M. J. (1981a). 16α-[^{77}Br]bromo-estradiol-17β: A high specific-activity, gamma-emitting tracer with uptake in rat uterus and induced mammary tumors. *J. Nucl. Med.* **22**:42–47.

Katzenellenbogen, J. A., Senderoff, S. F., Landvatter, S. W., McElvany, K. D., and Welch, M. J. (1981b). New halogenated estrogens with high binding affinity selectivity for the estrogen receptor: Potential breast tumor imaging agents. *J. Nucl. Med.* **22**:P20.

Katzenellenbogen, J. A., Senderoff, S. G., Welch, M. J., and McElvany, K. D. (1981c). [^{77}Br]-16α-bromoestradiol-17β, A high specific activity, gamma-emitting estrogen that shows selective, receptor-mediated uptake by uterus and DMBA-induced mammary tumors in rats. *J. Lab. Comp. Rad.* **XVIII**:81–83.

King, M. A., Casarett, G. W., Weber, D. A., Burgener, F. A., O'Mara, R. E., and Wilson, G. A. (1980). A study of irradiated bone III. Scintigraphic and radiographic detection of radiation-induced osteosarcomas. *J. Nucl. Med.* **21**:426–431.

Klinger, R., Georgi, P., and Wesch, H. (1976). Tierexperimentelle Untersuchungen zur positiven Tumordarstellung mit radiojodmarkierten Albumin-Makroaggregaten. *Nuklearmedizin* **XV**:91–94.

Kloss, G., and Leven, M. (1979). Accumulation of radioiodinated tyrosine derivatives in the adrenal medulla and in melanomas. *Eur. J. Nucl. Med.* **4**:179–186.

Kojima, S., Hama, Y., Sasaki, T., Osabe, W., and Kubodera, A. (1981). Analysis of the Radiorespirometric Pattern After Administration of [U-^{14}C]glucose to Ehrlich ascites tumor bearing mice. *Eur. J. Nucl. Med.* **6**:337–341.

Konikowski, T., Glenn, H. J., Haynie, T. P., and Jahns, M. F. (1977). Differences in [^{111}In] uptake in mouse brain sarcoma based on form of administration. *Int. J. Nucl. Med. Biol.* **4**:13–20.

Korn, N., Huang, C. C., Seevers, R. H., Rothwell, C., and Counsell, R. E. (1979).

Bisquaternary ammonium compounds as potential tumor imaging agents. *Int. J. Nucl. Med. Biol.* **6:**153–161.

Krohn, K. A., Meyers, J. M., DeNardo, G. L., and DeNardo, S. J. (1977). Comparison of radiolabeled bleomycins and gallium citrate in tumor-bearing mice. *J. Nucl. Med.* **18:**276–281.

Kruse, L. C., Reese, J. L., and Hart, L. K. (1979). Cancer pathophysiology, etiology and management. St. Louis: C. V. Mosby.

Larson, S. M., Rasey, J. S., and Allen, D. R. (1978). A transferrin mediated uptake of gallium-67 by EMT-6 sarcoma. *J. Nucl. Med.* **19:**715–716.

Larson, S. M., Grunbaum, Z., Rasey, J. S., Nelson, N. J., Allen, D. R., Harp, G. D., and Williams, D. L. (1979a). A transferrin binding macromolecular component of EMT-6 sarcoma (BALB/c mice). *J. Nucl. Med.* **20:**672.

Larson, S. M., Rasey, J. S., Allen, D. R., and Grunbaum, Z. (1979b). A transferrin-mediated uptake of gallium-67 by EMT-6 sarcoma II. Studies in vivo (BALB/c mice): Concise communication. *J. Nucl. Med.* **20:**843–846.

Larson, S. M., Rasey, J. S., Nelson, N. J., Grunbaum, Z., Allen, D. R., Harp, G. D., and Williams, D. L. (1979c). The kinetics of uptake and macromolecular binding of [^{67}Ga] and [^{59}Fe] by the EMT-6 sarcoma-like tumor of BALB/c mice. *Radiopharmaceuticals*, Vol. II. New York: Soc. Nucl. Med., pp. 297–308.

Larson, S. M., Rasey, J. S., Grunbaum, Z., and Allen, D. R. (1979d). Pharmacologic enhancement of gallium-67 tumor-to-blood ratios for EMT-6 sarcoma (BALB/c mice). *Radiology* **130:**241–244.

Larson, S. M., Grunbaum, Z., and Rasey, J. S. (1980a). Demonstration of gallium-67 bound to mouse transferrin in lysates of EMT-6 sarcoma cells harvested from BALB/c mice after prior intravenous injection of [^{67}Ga]-citrate. *J. Nucl. Med.* **21:**P23.

Larson, S. M., Grunbaum, Z., and Rasey, J. S. (1980b). Positron imaging feasibility studies. I. [^{3}H]-uridine and [^{14}C]-2-deoxyglucose. *J. Nucl. Med.* **21:**P32–33.

Larson, S. M., Grunbaum, Z., and Rasey, J. S. (1980c). Positron imaging feasibility studies: Selective tumor concentration of [^{3}H]-thymidine, [^{3}H]-uridine and [^{14}C]-2-deoxyglucose. *Radiology* **134:**771–773.

Larson, S. M., Weiden, P. L., Grunbaum, Z., Rasey, J. S., Kaplan, H. G., Graham, M. M., Harp, G. D., Sale, G. E., and Williams, D. L. (1981a). Positron imaging feasibility studies. I. Characteristics of [^{3}H] thymidine uptake in rodent and canine neoplasms: Concise communication. *J. Nucl. Med.* **22:**869–874.

Larson, S. M., Weiden, P. L., Grunbaum, Z., Kaplan, H. G., Rasey, J. S., Graham, M. M., Sale, G. E., Harp, G. D., and Williams, D. L. (1981b). Positron imaging feasibility studies. II: Characteristics of 2-deoxyglucose uptake in rodent and canine neoplasms: Concise communication. *J. Nucl. Med.* **22:**875–879.

Lee, Y. W., Knaus, E. E., and Wiebe, L. I. (1979a). Tumor uptake of radiolabeled pyrmidine bases and pyrimidine nucleosides. III. 2′-[^{82}Br]-bromo-2′-deoxyuridine. *Int. J. Nucl. Med. Biol.* **6:**109–112.

Lee, Y. W., Knaus, E. E., and Wiebe, L. I. (1979b). 2′-[^{82}Br]-bromo-2′-deoxyuridine for tumor uptake studies. *J. Lab. Comp. Rad.* **XVI:**167.

Levine, G., Ballou, B., Reiland, J., Soltner, D., Gumerman, L., and Hakala, T. (1980a). Localization of [^{131}I]-labeled tumor-specific monoclonal antibody in the tumor-bearing BALB/c mouse. *J. Nucl. Med.* **21:**570–573.

Levine, G., Ballou, B., Reiland, J., Solter, D., Gumerman, L. W., and Hakala, T.

(1980b). Tumor imaging in mice with [^{131}I] labeled monoclonal antibody. *J. Nucl. Med.* **21:**P54.

Levine, G., Ballou, B., Reiland, J., Solter, D., Gumerman, L., and Hakala, T. (1981). Tumor imaging with [^{131}I] monoclonal antibody and antibody fragment in mice. *J. Nucl. Med.* **22:**P35.

Lieberman, L. M., Young, D., Petering, D. H., and Minkel, D. (1976). Tissue distribution studies in mice with a [^{64}Cu] labeled thiosemicarbazone. *J. Nucl. Med.* **17:**568.

Louis, C. J. (1978). Tumors: Basic principles and clinical aspects. London: Churchill Livingstone.

Manger, R. J., Brown, L. E., Wieland, D. M., and Beierwaltes, W. H. (1981). Potential radiopharmaceuticals for diagnosis and treatment of carcinoid tumor. *J. Lab. Comp. Rad.* **XVIII:**45.

Markoe, A. M., Risch, V. R., Heindel, N. D., Emrich, J., Lippincott, W., and Brady, L. W. (1977). Tissue distribution of 5-thio-D-glucose in hamsters bearing pancreatic tumor models. *J. Nucl. Med.* **18:**610.

Markoe, A. M., Risch, V. R., Heindel, N. D., Emrich, J., Lippincott, W., Honda, T., and Brady, L. W. (1979). Biodistribution and pharmacokinetics of [^{35}S]-labelled 5-thio-D-glucose in hamsters bearing pancreatic tumors. *J. Nucl. Med.* **20:**753–760.

McDonald, J. M., Gelbard, A. S., Clarke, L. P., Christie, T. R., and Laughlin, J. S. (1976). Imaging of tumors involving bone with [^{13}N]-glutamic acid. *Radiology* **120:**623–626.

McElvany, K. D., Katzenellenbogen, J. A., Senderoff, S. G., Siegel, B. A., and Welch, M. J. (1981). Tissue distribution, clearance rates and radiation dosimetry of 16α[^{77}Br]-bromo-estradiol-17β. *J. Nucl. Med.* **22:**P19–20.

Mende, T., Wollny, G., Gens, J., and Schubert, J. (1979). Distribution and kinetic of iodine-labelled diaethylstilboestroldiphosphate (cytonal) after intravenous injection. *Eur. J. Nucl. Med.* **4:**133.

Miller, T. R., Siegel, B. A., Fair, W. R., Smith, E. K., and Welch, M. J. (1978). Imaging of canine tumors with [^{11}C]-methylputrescine. *Radiology* **129:**221–223.

Mitruka, B. M., Rawnsley, H. M., and Vadehra, D. V. (1976). *Animals for Medical Research.* New York: Wiley.

Muranaka, A., Ito, Y., Hashimoto, M., Namba, M., Nishitani, K., Otsuka, N., Nagai, K., Narabayashi, I., and Kaji, T. (1980). Uptake and excretion of [^{67}Ga]-citrate in malignant tumors and normal cells. *Eur. J. Nucl. Med.* **5:**31–37.

Noujaim, A. A., Lentle, B. C., Hill, J. R., Terner, U. K., and Wong, H. (1978). An *in vivo* model for gallium uptake studies in tumors. *J. Nucl. Med.* **19:**715.

Oster, Z. H., Som, P., Sacker, D. F., and Atkins, H. L. (1980). Time-related [^{67}Ga] citrate concentration in tumor and abscess and the effect of desferrioxamine (DFO). *J. Nucl. Med.* **21:**P24.

Otsuka, N., Ito, Y., Nagai, K., Terashima, H., Yanagimoto, S., and Muranaka, A. (1981). Effects of prostaglandin on experimental bone malignancy and on scintigrams of bone and marrow. *J. Nucl. Med.* **22:**433–440.

Packer, S., Lambrecht, R. M., Atkins, H. L., and Wolf, A. P. (1976). Short-lived radiopharmaceuticals for noncontact detection of ocular melanoma. *Nuclear Ophthalmology,* M. Croll, ed. New York: Wiley, pp. 112–121.

Packer, S., Lambrecht, R. M., Christman, D. R., Ansari, A. N., Wolf, A. P., and

Atkins, H. L. (1977). Metal isotopes used as radioactive indicators of ocular melanoma. *Am. J. Ophth.* **83:**80–94.

Pastakia, B., Lieberman, L. M., Gatly, S. J., Young, D., Petering, D. H., and Minkel, D. (1980). Tissue distribution of copper-labeled 3-ethoxy-2-oxobutyraldehyde bis(thiosemicarbazone)([^{64}Cu] KTS) in mice and rats: Concise communication. *J. Nucl. Med.* **21:**67–70.

Pertynski, T., Reznik, G., Mariss, P., Ketkar, M. B., Haindl, H., and Dieffenbach, G. (1979). The influence of chemotherapy on the distribution of [^{57}Co]-bleomycin in chemically induced squamous cell carcinoma of mouse skin. *Eur. J. Nucl. Med.* **4:**55–59.

Pettit, W. A., Tilbury, R. S., Digenis, G. A., Zedeck, M. S., Helson, L. R., Russ, G. R., and Mortara, R. H. (1976). Preparation of [^{13}N]BCNU and its tissue distribution in rats. *J. Nucl. Med.* **17:**558.

Priester, W. A., and Mantel, N. M. (1971). Occurrence of tumors in domestic animals. *J. Nat. Cancer Inst.* **47:**1333–1344.

Raban, P., Brousil, J., and Svihovcova, P. (1979). Chemical and biological properties of bleomycin labelled with [^{57}Co]. *Eur. J. Nucl. Med.* **4:**191–197.

Rayudu, G. V. S., Clark. P., Sukerkar, A. N., Ali, A., Turner, D. A., and Fordham, E. W. (1978). Sarcoma-180-tumor localization: In-111-oxine and [^{67}Ga]-oxine complexes versus [^{111}In]-chloride and [^{67}Ga]-citrate. *J. Nucl. Med.* **19:**716.

Rembelska, M., Liniecki, J., Pertynski, T., Oginski, M., Koniarek, B., and Hajdukiewicz, Z. (1979). Affinity of [^{58}Co]-bleomycin, [^{58}Co]-A_2-bleomycin and [^{58}Co]-citrate to experimental tumors in mice. *Eur. J. Nucl. Med.* **4:**137.

Reske, S. N., Vyska, K., Reske, K., and Feinendegen, L. E. (1980). I-131 DNase, a potential tumor imaging agent. *J. Nucl. Med.* **21:**P36.

Risch, V. R., Markoe, A. M., and Honda, T. (1978). *The Chemistry of Radiopharmaceuticals,* N. D. Heindel, H. D. Burns, T. Honda and L. W. Brady, eds. New York: Masson, Ch. 8.

Robbins, P. J., Sodd, V. J., Scholz, K. L., and Forman, D. L. (1976). Evaluation of antimony-177 for tumor imaging. *Int. J. Nucl. Med. Biol.* **3:**56–57.

Roberts, D. C. (1969). Transplanted tumors of rats and mice: An index of tumors and host strains. London: Laboratory Animals Ltd.

Robins, A. B., and Taylor, D. M. (1981). Iodine-123-iododeoxyuridine: A potential indicator of tumor response to treatment. *Int. J. Nucl. Med. Biol.* **8:**53–63.

Rosseau, J., Autenrieth, D., and van Lier, J. E. (1980). Technetium-99m tetrasulophthalocyanine as a potential tumor scanning agent. *J. Nucl. Med.* **21:**P37.

Rössler, K., Meyer, G.-J., and Stöcklin, G. (1979). Labelling and animal distribution studies of 5-astatouracil and 5-astatodeoxyuridine (211-At). *J. Lab. Comp. Rad.* **XIII:**271.

Russ, G. A., and Campione, M. A. (1981). Evaluation of radiolabeled 5-fluorouracils for nuclear scintigraphy. *J. Nucl. Med.* **22:**P50.

Sartiano, G. P., Levine, G., and Boggs, S. S. (1980). Influence of dimercaperol (2,3-dimercapto-1-propanol) on the pharmacokinetics of [^{111}In]bleomycin in the P-1798 lymphoma-bearing BALB/c mouse. *Oncology* **37:**120–124.

Schabel, F. M., Jr. (1975). Animal models as predictive systems. In *Cancer Chemotherapy: Fundamental Concepts and Recent Advances*. Chicago: Yearbook Medical Publishers, pp. 323–355.

Schumacher, J., Osswald, H., Wayss, K., and Maier-Borst, W. (1976). Uptake and excretion of [^{61}Cu] and [^{64}Cu]-myambutol by the liver. Experimental studies in rats and possibilities of a functional liver test in humans. *Int. J. Nucl. Med. Biol.* **3:**83–86.

Sephton, R. G., Hodgson, G. A., DeAbrew, S., and Harris, A. W. (1978). [^{67}Ga] and [^{59}Fe] distributions in mice. *J. Nucl. Med.* **19:**930–935.

Shani, J., Wolf, M., Schlesinger, T., Atkins, H. L., Bradley-Moore, P. R., Casella, V., Fowler, J. S., Greenberg, D., Ido, T., Lambrecht, R. M., MacGregor, R., Mantescu, C., Neirinckx, R., Som, P., and Wolf, A. P. (1978). Distribution of 18-F-5-fluorouracil in tumor-bearing mice and rats. *Int. J. Nucl. Med. Biol.* **5:**19–28.

Shani, J., Lieberman, L. M., Cais, M., Josephi, Y., and Shimoni, M. (1980). Synthesis of [^{203}Hg]-estradiol and its binding to mammary tumors. *J. Nucl. Med.* **21:**P37.

Shani, J., Lieberman, L. M., Cais, M., Josephi, Y., and Shimoni, M. (1981). Studies with [^{203}Hg]-estradiol for the detection of mammary tumors. *J. Lab. Comp. Rad.* **XVIII:**110–111.

Shaughnessy, W. J., Gatley, S. J., Simpkin, T. J., and Lieberman, L. M. (1980). Tissue distribution studies of 17-β(16α-[^{125}I])iodestradiol in rats bearing mammary adenocarcinoma. *J. Nucl. Med.* **21:**P36.

Shukla, S. K. (1979). Increasing the tumor specificity of [^{67}Ga]-radiopharmaceuticals. *J. Lab. Comp. Rad.* **XVI:**192.

Shukla, S. K., Castelli, L., Blotta, I., Delle Femmine, P., Caroli, S., and Alimonti, S. (1981). Role of citrate ions on the distribution of gallium in tumor-bearing rats and mice. *J. Lab. Comp. Rad.* **XVIIII:**215–216.

Som, P., Atkins, H. L., Bandyopadhyay, D., Fowler, J. S., McGregor, R. R., Matsui, K., Oster, Z. H., Sacker, D. F., Shiue, C.-Y., and Wolf, A. P. (1979). Early detection of neoplasms labeled with sugar analogue. *J. Nucl. Med.* **20:**662.

Som, P., Atkins, H. L., Bandyopadhyay, D., Fowler, J. S., McGregor, R. R., Matsui, K., Oster, Z. H., Sacker, D. F., Shiue, C. Y., Turner, H., Wan, C.-N., Wolf, A. P., and Zabinski, S. V. (1980). Fluorinated glucose analogue, 2-fluoro-2-deoxy-D-glucose [^{18}F]: Non-toxic tracer for rapid tumor detection. *J. Nucl. Med.* **21:**670–675.

Srivastava, S. C., Richards, P., Meinken, G. E., Pande, S. G., Som, P., Atkins, H. L., Larson, S. M., Grunbaum, Z., and Rasey, J. S. (1979a). Radiolabeled ruthenium compounds as potential tumor-localizing agents. *J. Nucl. Med.* **20:**684–685.

Srivastava, S. C., Richards, P., Meinken, G. E., Som, P., Atkins, H. L., Larson, S. M., Grunbaum, Z., Rasey, J. S., Dowling, M., and Clarke, M. J. (1979b). Evaluation of radiolabeled ruthenium compounds as tumor-localizing agents. *Radiopharmaceuticals,* Vol. II. New York: Soc. Nucl. Med., pp. 265–274.

Srivastava, S. C., Meinken, G. E., Richards, P., Larson, S. M., Grunbaum, Z., and Rasey, J. S. (1980). Ruthenium labeled transferrin—A potential tumor-localizing agent. *J. Nucl. Med.* **21:**79.

Sugiura, K., and Stock, C. C. (1955). Studies in a tumor spectrum III. *Cancer Res.* **15:**38–51.

Takeda, S., Uchida, T., and Matsuzawa, T. (1977). A comparative study on lysosomal accumulation of gallium-67 and indium-111 in Morris hepatoma 7316A. *J. Nucl. Med.* **18:**835–839.

Terner, U. K., Wong, H., Noujaim, A., Lentle, B. C., and Hill, J. R. (1979). Differential uptake of [^{67}Ga] in canine tumors and abscesses. *Radiopharmaceuticals,* Vol. II. New York: Soc. Nucl. Med., pp. 309–319.

Tse, J., Noujaim, A., and Wiebe, L. I. (1980). Studies on the use of samarium-153 radiopharmaceuticals in oncological diagnosis. *Eur. J. Nucl. Med.* **5:**A16.

Vallabhajosula, S. R., Harwig, J. F., and Wolf, W. (1979). Mechanism of tumor localization of gallium: Effect of tumor pH. *J. Nucl. Med.* **20:**655.

Vallabhajosula, S. R., Harwig, J. F., and Wolf, W. (1981). pH Dependent uptake of gallium by tumor cells *in vivo* and *in vitro:* Effect of glucose. *J. Lab. Comp. Rad.* **XVIII:**212–214.

van den Brom, W. E., Rijnberk, A., Lubberink, A. M. E., and de Bruijne, J. J. (1979). Uptake of [^{131}I]-19-cholesterol by normal and spontaneously hyperfunctioning canine adrenals. *Eur. J. Nucl. Med.* **4:**61–67.

van Langevelde, A., Beerling-van der Molen, H. D., Journee-de Korver, J. G., Paans, A. M. J., and Vaalburg, W. (1981). [^{11}C] labeled melanin precursors as radiopharmaceuticals for the detection of eye melanoma. *J. Nucl. Med.* **22:**76.

Volm, M., Gericke, D., Schuhmacher, J., and Wayss, K. (1977). Enhancement of incorporation of [^{131}I]-idododeoxyuridine into tumors after application of *Clostridium oncolyticum s. butyricum* (M55). *Eur. J. Nucl. Med.* **2:**117–120.

Vos, C. M., Westera, G., van der Jagt, P. J., and van Zanten, B. (1979). The effect of dose loading and of double labelling with [^{57}Co] and [^{125}I] on the tissue distribution in animals. *Eur. J. Nucl. Med.* **4:**393–396.

Washburn, L. C., Sun, T. T., Byrd, B. L., Hayes, R. L., and Butler, T. A. (1979). 1-Aminocyclobutane[^{11}C]carboxylic acid, a potential tumor-seeking agent. *J. Nucl. Med.* **20:**1055–1061.

Washburn, L. C., Sun, T. T., Byrd, B. L., Hayes, R. L., and Butler, T. A. (1979b). DL-[Carboxyl-[^{11}C]tryptophan, a potential agent for pancreatic imaging. Production and pre-clinical investigations. *J. Nucl. Med.* **20:**857–864.

Welch, M. J., Coleman, E., Straatman, M. G., Ashberry, B. E., Primeau, J. L., Fair, W. R., and Ter-Pogossian, M. M. (1977). Carbon-11-labelled methylated polyamine analogues: Uptake in prostate and tumor in animal models. *J. Nucl. Med.* **18:**74–78.

Welch, M. J., Coleman, E., Straatman, M. G., Ashberry, B. E., Primeau, J. L., Fair, W. R., and Ter-Pogossian, M. M. (1976). Carbon-11 labeled *N*-methyl 1,4-diaminobutane: A putrescine analog for prostate and tumor localization. *J. Nucl. Med.* **17:**525.

Wenzel, M., Nipper, E., and Klose, W. (1977). Biochemistry of metalocenes 1. Distribution of [^{59}Fe] or [^{103}Ru]-labelled metalocene carboxylic acid in mice. *J. Nucl. Med.* **18:**367–372.

Wenzel, M., Schneider, M., Bier, J., Benders, P., and Schachschneider, G. (1979). Cytostatika-Konzentrationen in Organen und Tumoren: Vergleich nach intravenoser und intratumoraler Injektion. *J. Cancer Res. Clin. Oncol.* **95:**147–157.

Wiebe, L. I., Lee, Y, Knaus, E., and Abrams, D. (1980). The synthesis of 6-radiohalogenopyrimidines for evaluation as radiopharmaceuticals in diagnostic oncology. *Eur. J. Nucl. Med.* **5:**A17.

Witcofski, R. L., and Pizzarello, D. J. (1976). Influence of dose rate on carcinogenesis resulting from X-ray, [^{113m}In], and [^{198}Au] irradiation. *J. Nucl. Med.* **17:**715–718.

Wolf, W., Shani, J., Yong, D., and Vine, E. (1977). Radiopharmacokinetics of antitumor agents: Fluorine-18 5-fluoruracil. *J. Nucl. Med.* **18:**617.

Wong, H., Tener, U. K., English, D., Noujaim, A. A., Lentle, B. C., and Hill, J. R. (1980). The role of transferrin in the *in vivo* uptake of gallium-67 in a canine tumor. *Int. J. Nucl. Med. Biol.* **7:**9–16.

Wortman, J., DeNardo, S., DeNardo, G., Huang, S.-M., Krohn, K., and Song, C. (1976). Thromboplastic activity (TA) and fibrinolytic activity (FA) of normal and neoplastic tissue. *J. Nucl. Med.* **17**:566–567.

Yeh, S. D. J., and Helson, L. (1978). Studies of tumor localizing radionuclides in transplanted human tumors in nude mice. *J. Nucl. Med.* **19**:716.

Yeh, S. D., Helson, L., and Grando, R. (1979). Tumor-localizing radionuclides heterotransplanted human tumors in nude mice. *Int. J. Nucl. Med. Biol.* **6**:169–173.

Yokoyama, A., Hata, N., Saji, H., Horiuchi, K., Tanaka, H., Morita, R., and Torizuka, K. (1981). Chemically designed [^{99m}Tc] radiopharmaceuticals for the tumor diagnosis. *J. Nucl. Med.* **22**:P69.

Zalutsky, M. R., Friedman, A. M., Sullivan, J. C., Ruby, S. L., and Rayudu, G. V. S. (1977). Mossbauer spectroscopic study of the chemical bonding of [^{57}Co]-bleomycin in tumors. *J. Nucl. Med.* **18**:602–603.

Zimmerman, M., and Hale, T. (1976). Verteilungsstudie mit [^{57}Co], [^{51}Cr], [^{111}In], und [^{99m}Tc] bleomycin in Mausen mit einem soliden Ehrlich-karzinom. *Nuklearmedizin* **XV**:176–182.

Zimmermann, M., and Schmutz, H. (1976). Radio diagnosis using [^{125}I] fibrinogen and [^{203}Hg] thymidine. *Eur. J. Nucl. Med.* **1**:251–254.

CHAPTER 5

^{111}In-Labeled Platelets for the Detection of Vascular Disorders in Animal Models

CARLA J. MATHIAS and MICHAEL J. WELCH

INTRODUCTION

Radiolabeled blood components were developed to better identify various parameters including: total blood volume (Wennesland et al., 1959; Radia et al., 1981), plasma volume (Hurley, 1974), cell kinetics (Heaton et al., 1979; Heyns et al., 1980; Pastakia, 1978) and the detection of abscess or thrombosis (Goodwin et al., 1978; Coleman et al., 1980; Thakur et al., 1977; Beswick et al., 1979; Krohn and Welch, 1977).

Isolated cellular components have been successively radiolabeled with [^{111}In]-8-hydroxyquinoline* (Scheffel et al., 1979; Thakur et al., 1976) and ^{51}Cr-sodium chromate (Aas and Gardner, 1958; Bjornson and Aursner, 1977; Lewis, S., 1977; Schmidt et al., 1982). Indium-111-labeled platelets have many advantages over ^{51}Cr-labeled platelets; these include high labeling efficiency with a high initial *in vivo* recovery, an appropriate radioisotope half-life ($t_{1/2}$ = 2.8 days) for measuring most cell survivals, and high-energy gamma emissions suitable for scintigraphic detection (174 KeV and 243 KeV) (Scheffel et al., 1977; Welch and Mathias 1980; Cunningham and Siegel, 1982; Born, 1980). ^{111}In-labeled platelets have been described by many investigators as a useful agent for scintigraphically detecting deep vein thrombosis (Davis and Welch, 1980b; French et al., 1981; Grimely et al., 1981; Knight et al., 1978), atherosclerotic lesions (Davis et al., 1978, 1980a;

* [^{111}In]acetylacetone (Mathias et al., 1981, Sinn and Silvester, 1979), [^{111}In]tropolone (Dewanjee et al., 1981, Hendershott et al., 1982; Danpure et al., 1982) and other chelating agents which form a lipophilic complex (McAfee and Thakur, 1976) have also been used to label blood cells.

Powers et al., 1982a), mural thrombi (Davis and Welch, 1979) and coronary artery thrombi (Bergmann et al., 1982b; Ezekowitz et al., 1981a).

Unfortunately, minute platelet accumulation or monolayer adhesion on thrombogenic surfaces is not readily imaged even 24 hr after administration of ^{111}In-labeled platelets (Riba et al., 1979; Powers et al., 1982b). Presumably, the concentration of radiolabeled platelets deposited on the lesion is increased relative to the amount of radioactivity circulating in the blood. To best define small platelet deposition, a means for determining the normal blood pool is essential. Red blood cells (RBC) labeled with ^{99m}Tc have been used to scintigraphically localize normal blood pool areas (Schwartz, 1977); and with computer assisted collection of serial images this dual isotope technique can effectively enhance the sensitivity and specificity of images obtained with ^{111}In-labeled platelets (Powers et al., 1982c; Bergmann et al., 1982b). A noninvasive technique that accurately measures the intravascular platelet deposition is necessary so that qualitative interpretation of images may be eliminated (Sutherland et al., 1982).

During the hemostatic process, several reactions are observed which progress from vasoconstriction, platelet adhesion and aggregation to collagen fibrils (Jamieson, 1974; Walsh, 1977); to clot reinforcement by fibrin (Niewiarowski, 1972, 1978; Mustard et al., 1977b), followed by fibrinolysis (Hedner and Nilsson, 1981; Vermylen, 1978; Astrup, 1969; Welch, 1899). Each of these phenomena yields numerous factors via the coagulation pathway (MacFarlane, 1972), the vascular surface (i.e., stimulation of prostaglandin production) (Moncada and Vane, 1979; Tsao, 1979), and fibrinolysis (stimulation of cAMP) (Nalbandian and Henry, 1978). These factors react with or cause an alteration of the circulating platelets (Woolf, 1978). Other platelet–vascular-related diseases (i.e., stroke, hypertension, myocardial infarction and atherosclerosis) have been described as being the result of a platelet (or platelet-associated factor) mediated process; this means that the role of prostaglandins, proteoglycans, cAMP, calcium, and lysosomal enzymes is perhaps more prevalent than yet demonstrated (Mustard, 1977a; Sherry, 1977).

Many human diseases have been modeled in several animal species. Some important species variations are found in platelet response to various aggregation stimulating agents (Addonizio et al., 1978; Sinakos and Caen, 1967), platelet adhesion (Mills, 1970), effects of platelet inhibitors and other drugs (Packham and Mustard, 1977; Russell et al., 1966; Vane and Moncada, 1980), anticoagulant effects (Cliffton and Downee, 1950; Mills, 1970), platelet size and the involvement at areas of thrombosis (Guest et al., 1948; Hawkey, 1974; Mustard, 1976; Packham et al., 1968), or other thrombogenic surfaces (French et al., 1964; Roohk et al., 1976; Schafer and Handen, 1979). There is tremendous variation in platelet counts and size. For example, the volume of a cat platelet is large (12.2 μm^3) compared to man, which is 5.8 μm^3, and rabbit (4.1 μm^3) (Dodds, 1978). The circulating platelet count in animals is generally higher than that of man (Table 5-1).

TABLE 5-1. Species Variation of Platelet Aggregation Responses[a]

Species	*Circulating Platelet Count*	in vitro *Platelet Aggregation Stimulant*				
		ADP	*con*	*TH*	*Serotonin*	*Adrendin*
Man	210 K	B	M	B	M	B
Monkey	260 K	B^x	M	M	M	B^x
Baboon		B	M	M	M	B
Rabbit		M	M	M	M	—
Guinea Pig		B^x	M	M	—	
Rat		M	M	M	M	—
Cat		B		B	B	B
Dog	320 K	B^x	M	M	M	B^x
Pig		M		M		—
Sheep		M		M	M	—
Goat		M	M	M		—
Cow		M	M	M		
Horse		B^x	M	M	B^{x}P	—
Mouse		M		M		—
Hamster		M	M	M		
Bird and Reptile		—	—	M	—	—

[a] Data from Dobbs, 1978; Hawkey, 1970; MacMillan and Sim, 1970.
[b] B, biphasic; B^x, biphasic occasionally; M, monophasic;—, no response.

Consideration should be given use of anticoagulants; an anticoagulant is essential throughout *in vitro* platelet procedures and some, such as heparin and ethylenediaminetetraacetic acid (EDTA), can cause suboptimal results in rabbits, dogs, and cats, depending somewhat on the concentration (Scrutton and Egan, 1979; Aster, 1971). Sodium citrate solution (3.8%, pH 7.35) seems more optimal at 1 : 9 (v : v) concentration (Karpatkin et al., 1970; Kien et al., 1971). Some species, however, require less sodium citrate (3% in rats) in order to maintain *in vitro* platelet aggregation response to adenosine diphosphate (ADP) (Dodds, 1978).

Platelet viability has been evaluated with the use of *in vitro* platelet aggregation studies (Born et al., 1964; Mustard and Packham, 1970; Weiss, 1975). The measurement of the response to various aggregation stimulating agents (ADP, epinephrine, collagen, arachidonic acid, and thrombin) has been described previously (Welch and Mathias, 1980; Born, 1974; Holmsen, 1974; Roper-Drewinko et al., 1981). The turbidometric technique employed is based on the fact that the formation of aggregation in a stirred suspension of platelets results in an increase in the transmission of light through the solution (Weiss, 1974) (Fig. 5-1).

Platelet function, involved in the coagulation mechanism and formation of the hemostatic plug, may be described in a sequence of reactions: (1) shape change, (2) adhesion reaction, (3) primary aggregation, (4) release reaction, and (5) secondary aggregation. Immediately following the interaction of an

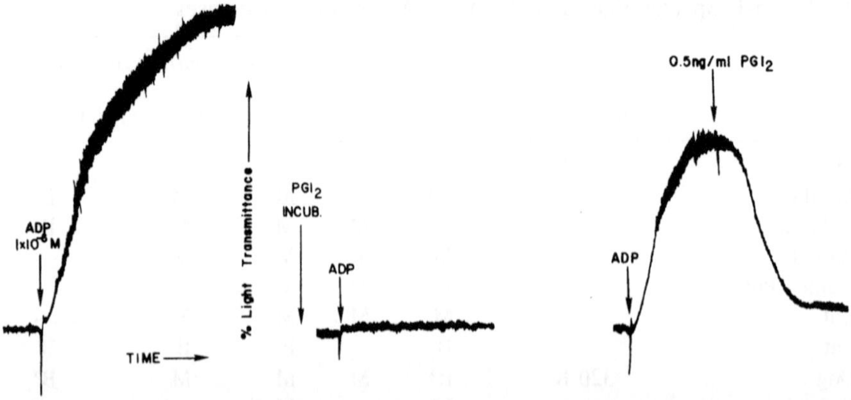

FIG. 5-1. (a) Normal platelet aggregation response to a relatively high concentration of ADP (1×10^{-5} *M*) (left). The decrease after the aggregation stimulus (platelet shape change) followed by a rapid increase in light transmission (primary and secondary phase aggregation). (b) If platelet rich plasma (PRP) is incubated with 0.5 ng/ml PGI_2 and then a platelet aggregation stimulating agent is added, there is no change in the light transmitted (i.e., platelet aggregation is inhibited) (center). (c) Finally, reversal of platelet aggregation in response to ADP (1×10^{-5} *M*) by 0.5 ng/ml of PGI_2 1 min after ADP was added (right).

aggregation stimulus with the platelet membrane the platelets change from ellipsoids to spiny spheres, probably not accompanied by any change in cell volume. The initial reaction is not calcium dependent and is reversible at very low stimulus levels. With normal concentrations of stimuli the shape change is followed by an adhesion reaction. This is the process of platelet adherence to surfaces other than platelets. Platelet adhesion is distinct from aggregation because aggregation refers to adherence of platelets to each other. Platelet aggregation, primary phase, is the adherence of platelets to each other and is accompanied by a viscous metamorphous and a release reaction. It can be induced *in vitro* under the correct conditions; it occurs *in vivo* as a result of various stimuli. Optimum conditions include pH 6.8–8.5, temperature of 37°C, and constant stirring and the appropriate aggregation agent. The release reaction refers to the biochemical reactions occurring when adenine nucleotides, serotonin proteins, and amino acids are extruded rapidly from platelet alpha granules and dense bodies to extracellular medium. The alpha granules contain lysomal enzymes and dense bodies contain amines and nucleotides (e.g. calcium, ATP, ADP, and serotonin) (Holmsen et al., 1969). ADP, adrenaline, or low concentrations of thrombin or collagen can induce release of dense body constituents not accompanied by release of alpha granule constituents. However, release of both granules constituents can be induced by high concentrations of thrombin and collagen. Extracellular calcium is essential for the release reaction that accompanies aggregation (Holmsen, 1974). The release of ADP during the release reaction causes a second wave of aggregation, which causes further release, which results in a cumulative effect of total aggregation. In some cases the second wave of aggregation is not discernible, the aggregation response appearing as a continuous reaction since the stimulus addition to the total aggregation.

There are obvious similarities between the *in vivo* platelet aggregation responses of man and nonhuman primates. Dogs have a similar pattern of aggregation responses but require a higher concentration of stimulating agents (ADP and adrenalin) to obtain biphasic hyper-responses like those of man (Dodds, 1978; Harris et al., 1981). Cats are unusual in that they have a biphasic response to most stimulating agents (MacMillan and Sim, 1970). The other relevant species have only monophasic aggregation responses to most agents.

Prostaglandin endoperoxides and nonprostanoate derivatives of arachadonic acid may be the chemical initiator for the release reaction (Hamberg et al., 1975; Smith et al., 1977). Apparently, aspirin (which inhibits prostaglandin synthesis) (Roth et al., 1975) inhibits dense body release induced by ADP, adrenaline, or low concentration of collagen but has no effect on alpha granule release or on dense body release induced by high concentrations of collagen or thrombin (Smith and Willis, 1971). It is important for comparative experiments that the platelet concentration, pH, temperatures, and anticoagulant remain constant (Tang and Frozinovic, 1977; Constantine, 1965).

Plasminogen also influences platelet function *in vivo* (Mysliwiec et al., 1980). Plasminogen levels have been correlated to fibrinolytic potentials (Wautier and Coen, 1979; Ogston and Bennett, 1978). Relative to humans, pigs have less plasminogen and thereby a lesser fibrinolytic potential; rats and rabbits have increased (×2) plasminogen levels, and dogs have even greater plasminogen levels (×3). Monkeys, however, demonstrate a plasminogen level only slightly greater than that of humans (the concentration range includes that of humans) and most likely approximate the fibrinolytic potential of the human system (Gallimore et al., 1965; Hawkey, 1970; Mason and Read, 1971; Doolittle and Omcley, 1962; Niewiarowski and Latallo, 1959).

The accumulation of platelets at sites of vascular injury has been confirmed by techniques including electrical and mechanical injury (Branemark and Ekholm, 1968; Barnhart and Chen, 1978; Lewis and Westwick, 1977; French, 1969; Hirsch and Loewe, 1946; Grossman et al., 1978). Since 1951 the electrocoagulation method has been utilized to investigate thrombosis on mesenteric arteries, jugular veins, femoral veins, carotid arteries, femoral arteries, and coronary arteries (Lutz et al., 1951; Cowan and Monkhouse, 1966; Ashford and Freiman, 1968). This technique is reproducible in many species of animals and has been employed to evaluate the efficacy of antithrombotic agents; species differences have been reported; however, the differences are probably related to vessel size rather than the species (Brinkhous, 1978a; Didisheim, 1976). The histology obtained from the site of electrical injury indicates that the thrombus is the result of endothelium destruction and exposure of the internal elastic lamina and media (Duval et al., 1970; Baumgartner and Haudenschild, 1972). However, many investigators disagree, suggesting that platelets may adhere to sites of minute endothelial injury without cell wall breaks (Johnson, 1968). Platelets have been shown to adhere to other vascular components, such as basement membrane (which contains a collagenlike material) (Kelfalides and Wingler, 1966), elastin fibers (Ross and Bornstein, 1969) (present in large arteries, which rarely contain collagen fibers) (Hughes, 1962), and subendothelial collagen (Baumgartner and Spaet, 1970; Baumgartner et al., 1976, 1977a, 1977b; Mehta and Mehta, 1981; Zucker, 1980). It has been postulated that the access of plasma proteins to the subendothelium stimulates the evolution of thrombin, thereby inducing platelet aggregation (Majno and Palade, 1961).

For some time the electrical injury technique was used to produce thrombi *in vivo* as a reliable system for radiotracer modeling (Cowan and Monkhouse, 1966; Williams and Carey, 1959). These studies were primarily carried out in the femoral vein of dogs to evaluate thrombus imaging agents such as radioiodinated fibrin (Harwig et al., 1976; Coleman et al., 1975a), [^{125}I]-fibrinogen (Welch and Krohn, 1975), [^{123}I]-plasmin (Takeda and Nakabayashi, 1974), [^{123}I]- and [^{99m}Tc]-urokinase (Rhodes et al., 1973), [^{123}I]- and [^{99m}Tc]-streptokinase (Persson and Kemper, 1977), [^{123}I]-plasminogen (Harwig et al., 1977), and [^{111}In]platelets for deep venous thrombi (DVT) (Davis and Welch, 1979; Welch et al., 1980b). [^{111}In]-platelets and other agents were compared and results demonstrate that platelet deposition in these thrombi is extensive (White and Heptinstall, 1978; Knight et al., 1978). To better define the role of platelets and predict a useful treatment for DVT and thromboembolism, the electric injury model was used to produce DVT in canine femoral veins (Coleman et al., 1975b; Sawyer and Pate, 1953; Schwartz, 1962). ^{111}In-labeled platelets were infused sometime after the clot formation, and scintigraphic identification of the platelet accumulation was possible. In dogs maximum platelet deposition was observed approximately 4 hours after injection; however, in monkeys the clot-to-blood ratios continued to increase for about 24 hr. The more rapid fibrinolytic response in canines may explain the more rapid accumulation followed by clot dissolution (Moser et al., 1973); primates have a somewhat slower lysis response and thereby the thrombus would continue to accumulate platelets during that initial period (Fig. 5-2). The platelet deposition was a function not necessarily of clot size but of the degree of vascular damage. ^{111}In-labeled platelets have also been used to detect pulmonary emboli (PE) (McIlmoyle et al., 1977; Moser, 1980). The scintigraphic localization of PE in canines was reliable until 24 hr after embolization; beyond this time very little evidence of PE existed. This method of electrical injury, although consistent, introduces an element that is difficult to distinguish: platelet deposition in propagating thrombi from interaction on the injured vessel wall. To investigate the role of platelets in thrombosis, many other mechanisms need to be defined, such as the role of platelets in acute arterial thrombosis on denuded endothelium (Cotran, 1969; Sheppard and French, 1971), the proliferation of atherosclerotic plaques (Friedman and Byers, 1961; Harker et al., 1978), and white thrombus formation without vessel injury (Welch, 1887; Lutz et al., 1951).

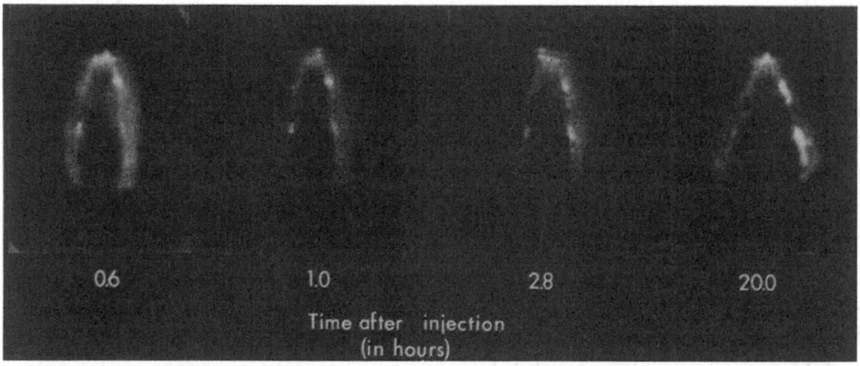

FIG. 5-2. Scintigrams of [^{111}In]-platelets obtained after injection in a nonhuman primate. The increased accumulation at the occlusive deep venous thrombi (produced by electroclotography) is more intense at 20 hr.

Four animal models with potential regions of platelet deposition have been evaluated using the dual isotope technique with ^{111}In-labeled platelets and ^{99m}Tc-labeled RBC. (1) An acute arterial endothelium injury in the abdominal aorta of nonhuman primates was examined 24 hr after ^{111}In-labeled platelets were administered. (2) A group of nonhuman primates with chronic atherogenic lesions was examined for platelet interaction with the atherosclerotic plaques in the abdominal aorta. (3) A canine model for acute coronary artery thrombosis was used to evaluate a platelet accumulation in an area with a large amount of circulating radioactive blood. (4) A canine model to evaluate platelet deposition on small-diameter vascular grafts was examined.

The calculations to determine the amount of radiolabeled platelets deposited on the region involved is slightly different in each case, since each model has intrinsic variations that are described later. The technical suitability of an animal is evaluated for vascular access, response to drugs (including anesthetics), and ability to tolerate the experiment.

ACUTE ARTERIAL LESION MODEL

There have been several different approaches to evaluate the platelet–endothelium interaction after an acute endothelial injury (Groves et al., 1979; Ramirez et al., 1981; Stemerman and Ross, 1972; Bourgain et al., 1979b; Dejana et al., 1980; Finkelstein et al., 1982). After a single injury in arteries of normal rabbits, a monolayer of platelets accumulate on the injured surface (Mustard et al., 1977c) and may lead to more severe thrombosis or may dissipate after several days. This is, of course, related to the rate of fibrinolysis and the rate of plasmin formation. Prostacyclin (PGI_2) is a potent smooth muscle relaxant and inhibitor of platelet function and thereby a potentially effective agent for the treatment of thrombosis (Moncada et al., 1976;

Gryglewski et al., 1978a). Some studies have demonstrated the inhibition of platelet accumulation in areas of vascular damage and exposed subendothelium (Adelman et al., 1981; Weiss and Turitto, 1979; Moncada and Vane, 1978; Higgs et al., 1978). The PGI_2 dose response varies from species to species, and some secondary effects have been documented (Creasy et al., 1979). Many species of animals have been used to examine the effects of PGI_2 on experimentally induced vascular lesions (Karim and Adarkan, 1979; Sinzinger et al., 1980; Bourgain, 1979a). Even though scintigraphic results may suggest platelet accumulation, *in vitro* confirmation by counting excised vascular specimens has thus far been necessary (Wu, 1981; Buchanan et al., 1981; Jaeger and Berntsen, 1979). Nonhuman primates (*Macaca fascicularis* and *Macaca nemestrina*) exhibit a similar coagulation system and *in vitro* platelet aggregation response as in the human situation. *In vitro* platelet aggregation inhibition and aggregation reversal is observed in both human and nonhuman primates (Fig. 5-1b, c) (Gryglewski et al., 1978b; Karim and Adarkan, 1979). These nonhuman primates could also tolerate the anesthesia necessary for long experimental procedures. ^{99m}Tc-labeled RBC were infused 24 hr after the administration of ^{111}In-labeled autologous platelets and image collection was carried out. Alternating 1-min images for the 243 keV photopeak of ^{111}In and the 140 keV photopeak of ^{99m}Tc were collected sequentially, digitized, and stored as a 64 × 64 matrix on a computer. After endothelium damage produced by an inflated balloon catheter, rapid, marked platelet deposition occurs (Fig. 5-3). Sufficient differences between undamaged artery and damaged artery were not readily detectable without increasing the imaging sensitivity with blood pool subtraction. Time–activity curves were generated from stored, sequential 1-min images that included a baseline obtained before damage, followed by a period before, during, and

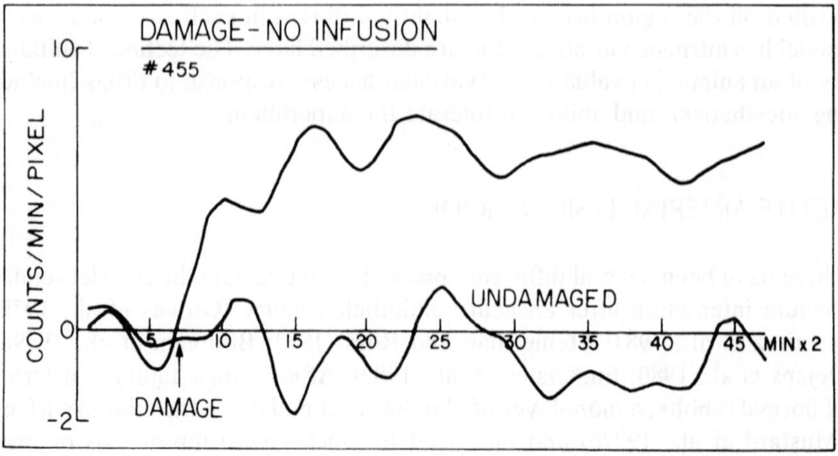

FIG. 5-3. Time–activity curves of baseline followed by the period after an acute arterial lesion formation in both a control aorta region (undamaged) and the damaged region.

TABLE 5-2. Inhibition of ADP-Induced Aggregation in Various Species by PGI_2[a]

Species	*Relative Potency* *PGI_2*	*Concentration* *PGI_2 (nM)*
Sheep	1	26.4
Horse	1.27	25.2
Rat	1.55	7.5
Rabbit	2.32	6.9
Baboon	5.88	4.4
Human	6.58	4.4
	7.0	3.6

[a] Data from Karim, 1977; Moncada, 1977.

after drug administration. The curves reflect platelet deposition on exposed subendothelium and reversal of mural platelet deposition by prostacyclin. It was found that PGI_2 can reverse platelet deposition on denuded endothelium but that the reversal was dependent on the time elapsed since subendothelium was exposed (Gryglewski et al., 1978b; Welch et al., 1981a). These results suggest that within 46 min after the arterial damage PGI_2 causes a decrease in platelet deposition, but after 46 min, PGI_2 was not significantly effective in decreasing the platelet deposition. These results were obtained infusing PGI_2 at 100 ng/kg min for 20 min, which did not alter the blood pressure or heart rate. Rabbits, however, are more resistant and can tolerate 1000 ng/kg min without hypotensive effects (Armstrong et al., 1978; Bergman et al., 1981; Szczeklik et al., 1980). Humans are more sensitive and could tolerate PGI_2 at only 10 ng/kg min before hypotension was induced (Szczeklik et al., 1978; Fitzgerald et al., 1979) (Table 5-2).

Platelet accumulation in the damaged region is rapid, whereas activity in the undamaged areas remains relatively stable (Fig. 5-3). Even though the scintigraphic images obtained immediately after subendothelial exposure and those obtained after PGI_2 infusion are not markedly different, the curves obtained by calculating the radioactivity associated with platelet deposition are dramatic. The somewhat ambiguous scintigraphic information is clarified by this dual isotope technique; platelet accumulation and reversal are readily identified on the time–activity curves based on increased residual counts over the region of interest as compared to undamaged segments.

ATHEROSCLEROSIS

Hypercholesterolima has been shown to cause severe endothelial injury (Ross and Harker, 1976; Pearson, et al., 1977). In response to endothelial cell injury and focal desquamation, platelets accumulate and aggregate, during which they release the granule constituents (Ross, 1979). These products are believed to stimulate smooth muscle cell proliferation and a change in the vessel wall permeability. This is followed by the formation of connective tissue matrix and lipid accumulation (Bullock et al., 1975; Harker, 1979). Pharmacologic inhibition of platelet function is an important approach to the prevention of atherogenesis since the platelet release is thought to be an initiating factor in the proliferation of smooth muscle cells, and thereby arteriosclerotic disease (Harker et al., 1978; Hornstra and Haddeman, 1977; Small, 1977; Kottke and Subbiah, 1978; Steinberg, 1979).

Experimental atherosclerosis has been induced in a number of different animal species (DePalma et al., 1972; Brinkhous and Bowie, 1978b; Clarkson, 1972), including rabbits (Cohen and McCombs, 1967; Moore and Inatocoycz, 1977; Spaet, 1977; Groves, 1979), dogs (Malmros and Sternby, 1968), pigeons (Clarkson et al., 1959; Lewis, J. C. 1977; Anitschkow and Chalatow, 1913), swine (Nam et al., 1973; Gottlieb and Lalich, 1954), and nonhuman primates (Wissler and Vesselinovitch, 1977; Stary and Malenow, 1982; Pick et al., 1974; Stemerman and Ross, 1972). Although various techniques have been employed to accelerate the onset of arterial plaques, the most obvious and representative animal model is diet-induced atheroma in nonhuman primates (Wagner and Clarkson, 1975; Kramsch and Hollander, 1968; Wu et al., 1975; Malinow, 1980; Honour et al., 1978). *Macaca aractoides* that had been on atherogenic diet for 4 years were studied. The high cholesterol diet (primarily egg yolks and lard) was maintained and cholesterol and triglyceride levels measured. The total cholesterol was elevated to greater than four times as much as baseline. The control animals were fed an identical diet without cholesterol (casein substituted). Although a few investigators have used angiography to document atherogenesis in monkeys (DePalma, 1972; Kramsch, 1968), there are many limitations for humans (Barndt et al., 1977), presumably making the difficulties greater in monkeys. An approach to the elucidation of the role of platelets in the progression of atherosclerosis would be to monitor platelets on atherosclerotic surfaces *in vivo* and partially evalute pharmaco-intervention (Small, 1977).

The animals were studied using ^{111}In-labeled autologous platelets and [^{99m}Tc]-RBCs (Powers et al., 1982a; Welch et al., 1981b). The primate was anesthetized, restrained, and positioned under a large field of view gamma camera so that the imaging field included heart, liver, spleen, abdominal aorta, and iliac bifurcation. Immediately after the ^{111}In-labeled platelets were administered, an image of both the 174 and 243 keV photopeaks of ^{111}In was collected (200–500 K counts) and stored on computer. The ^{99m}Tc-labeled RBC were administered and another image was obtained using the 243 keV photopeak of ^{111}In, allowing time for equilibration of the RBCs. An image utilizing the 140 keV photopeak of ^{99m}Tc was obtained while the animal remained in the same position. The imaging procedure was repeated 24 hr later after a second dose of [^{99m}Tc]-RBC was administered.

The unprocessed scintigraphic images obtained were unremarkable and showed no obvious foci of platelet deposition (Fig. 5-4); however, one might

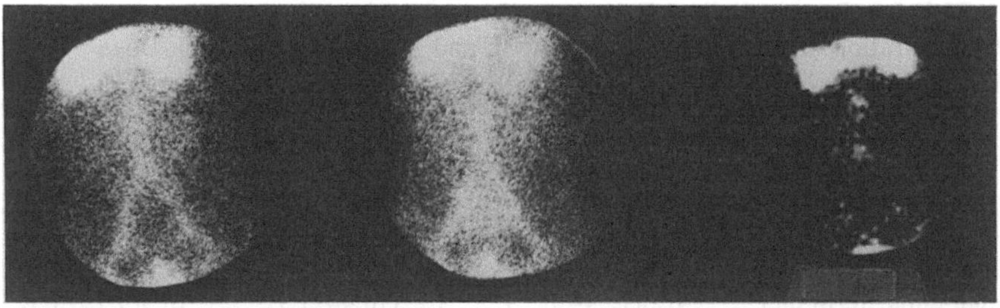

FIG. 5-4. Scintigrams obtained 24 hr after [^{111}In]-platelets were injected (left) and immediately after [^{99m}Tc]-RBC were administered (center). No obvious foci of platelet deposition is detected in either of these scintigrams, however, the computer-assisted subtracted imaged (right) represents defuse areas along the course of the abdominal aorta with several foci of increased intensity.

visualize persistent foci along the course of the abdominal aorta in the processed image, which represents only noncirculating ^{111}In activity or platelet depositon (% IE).

Unfortunately, this subtraction technique requires the identification of a normal blood pool within the field of view. In animals with known atherosclerotic lesions, it would be difficult to define a normal blood pool area since diffuse platelet adhesion/aggregation should continue in most areas. The processing technique was modified so that the platelet deposition (% IE) or percent indium excess could be calculated for each four-pixel-wide horizontal row within the limits of the abdominal aorta. These values were determined for both ^{111}In and ^{99m}Tc and the ratio for each corresponding row calculated. The calculation to determine % IE is

$$\%\ \text{IE} = \frac{^{111}\text{In counts in region of interest} - {}^{111}\text{In counts in blood pool}}{^{111}\text{In counts in blood pool}} \times 100$$

(Eq. 5-1)

Ideally, the reference ratio that should be selected is the ^{111}In to ^{99m}Tc ratio in a vascular area with no abnormal platelet deposition. Since that region is difficult to define in this model, the reference ratio was determined as the minimum ^{111}In to ^{99m}Tc ratio within the region of interest. Even if diffuse platelet deposition in the abdominal aorta exists, the minimum ^{111}In to ^{99m}Tc ratio will be greater than the true reference ratio. These assumptions, then, limit the sensitivity of this method for detecting platelet deposition. Profiles can be generated with % IE values that represent the platelet deposition along the abdominal aorta, but mean values for % IE for each row were determined and used for further analysis. Statistical analysis was performed using the Mann Whitney U test.

The animals were studied and the results demonstrate no significant difference in the mean % IE values between the control and diet animals;

TABLE 5-3. Percent Indium Excess (% IE) in the Abdominal Aorta 24 hr After Administration of ^{111}In-Platelets Untreated, and After Low- and High-Dose Aspirin

	Mean % IE 24 Hr After Injection		
	Untreated	*Low-Dose ASA*	*High-Dose ASA*
Control diet	15.8(9.9–18.5)	18.7(14.5–22.9)	11.7(8.4–14.9)
High-cholesterol diet	41.5(15.4–92.5)	16.2(10.4–21.7)	16.9(9.7–23.3)

[a] Data from Welch 1981b.

however, 24 hr after the radiolabeled platelets were injected, the diet animals had significantly greater amounts of platelet deposition in the abdominal aorta than the controls.

These same groups of animals were studied after (1) low dose (3.0 mg/kg day) aspirin or (2) high dose (30 mg/kg day) aspirin was administered for 5 consecutive days prior to the study (Pick et al., 1978). The results indicate that, in fact, either 3 mg/kg day or 30 mg/kg day of aspirin can inhibit platelet deposition on atherosclerotic plaques both immediately and 24 hr after the radiolabeled platelets were injected (Table 5-3) (Hauss et al., 1978). The platelet deposition (or % IE) is not uniform through the course of the abdominal aorta, which is consistent with the pathological findings (Bullock et al., 1975). The scintigraphic images showed no abnormal foci of accumulation along the course of the abdominal aorta, and there were no ulcerated or otherwise unusually thrombogenic plaques in the abdominal aortas of the diet animals on histological analysis.

CORONARY ARTERY THROMBOSIS

This same dual isotope technique has been successfully utilized in a canine model of coronary artery thrombosis (Bergmann et al., 1982c). Because the dog has a rapid fibrinolytic system relative to human, it probably does not closely mimic the clinical situation; however, nonhuman primates, which have similar fibrinolytic systems, were not justified, since these were terminal experiments. In fact, the canine model may provide an accelerated system for an otherwise lengthy observation period. The subtraction technique allows a relatively small thrombus (avg. 44 ± 28 mg) to be visualized even in an area with a large amount of blood pool, such as the heart. Intracoronary artery thrombosis occurs early after transmural myocardial infarction (DeWood et al., 1980) and has been identified as a site of rapid fibrinogen turnover (Erhardt et al., 1973; Mehta and Mehta, 1979; Moschos et al., 1974). Other investigators have shown that ^{111}In-labeled platelets can be used alone for visualization of coronary artery thrombi (Ezekowitz et al., 1981b) and interventricular thrombi (Riba et al., 1979; Stratton et al., 1981), but the relatively large blood pool within the area of interest interfered with the immediate localization of the platelet accumulation (Ezekowitz et al., 1981a). Previously, the coronary artery thrombus was induced by electroclotography and thereby was not a model of exclusively vascular thrombus, but also vascular wall damage and prostaglandin introduction (Salazar, 1961).

The coronary artery thrombus was induced by placing a small (5–7.5 mm in length) copper wire coil in the left anterior descending coronary artery (Kordenat and Kizidi, 1972). The thrombus occluded the artery 5–15 min after removing the guide wire. The autologous ^{111}In-labeled platelets and ^{99m}Tc-labeled RBC were prepared and injected either before the clot induction or 1 hr after clot formation as determined by arteriography. Alternating sequential images were collected, digitized, and stored on computer for processing. In the model, the origin of the right carotid artery was used as the reference or normal blood pool area. Representative subtracted images were produced by applying a subtraction algorithm corrected by the reference ratio to the appropriate pair of images. The scintigraphic images obtained from either ^{111}In or ^{99m}Tc were unremarkable; however, the subtracted image, even in the immediate situation, was suitable to localize an abnormal focal accumulation corresponding to the location of the coil by arteriography (Fig. 5-5).

The values for % IE in the region of the clot were determined and the *in vitro* clot to blood ratio was calculated (Table 5-4). The values obtained for clot weight to blood ratio and % IE in either case (radiolabeled cells injected either before or after clot formation) are similar; even though the values obtained when the [^{111}In]-platelets are injected after clot formation appear reduced, they are not statistically significant. A linear correlation exists of *in vitro* counts in the clot and *in vivo* % IE.

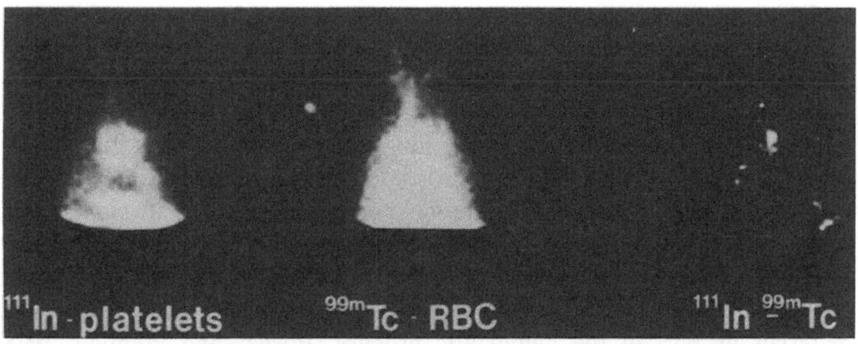

FIG. 5-5. Anterior views of the canine including the heart and liver with the use of [^{111}In]-platelets (left) and [^{99m}Tc]-RBC (center) 1 hr after administration. The subtracted image (right) represents platelet deposition in the area corresponding to the coil location as determined by arteriography.

TABLE 5-4. Evaluation of Streptokinase for Coronary Artery Thrombus Dissolution Using ^{111}In-Labeled Platelets

Time (hr) Cells Injected	*n*	*Clot-to-Blood Ratio*	*Clot Wt. (mg)*	*IE (%)*
−1	7	38.1 ± 14.1	46.8 ± 29.8	35.4 ± 24.6
1	5	27.5 ± 21.7	32.9 ± 24.8	20.7 ± 12.8
6	2	25.1 ± 2.3	27.6 ± 10.1	17.2 ± 1.2
−1 SK	3	22.1 ± 33.4	11.0 ± 3.4	16.8 ± 4.2
1 SK	3	9.1 ± 1.9	5.1 ± 4.2	10.1 ± 9.5

This model can be utilized to demonstrate the efficacy of thrombolytic therapy such as plasminogen inhibitors streptokinase and urokinase (Duckert, 1979; Fletcher and Sherry, 1960; Barlow, 1979; Rentrop et al., 1981; Bergmann et al., 1982a). Radiolabeled cells were administered either before or after a clot formation, as confirmed by arteriography and processed scintigrams. Image pairs are collected for a period of time thereafter (3 hr); then the intracoronary infusion of the drug is begun. Thrombolysis occurs approximately 45 min after the beginning of therapy with streptokinase (4000 U/min) marked by EKG reperfusion dysrhythmia (Cowley and Gold, 1982; Verstraete, 1978). More scintigraphic images are collected, and stored for processing. The dog has a repeat arteriogram and then is sacrificed. The excised heart is examined and specimens, including the reperfused tissue, coil, artery surrounding the coil, infarcted tissue, and blood, are obtained for *in vitro* counting.

The thrombolysis with streptokinase was successful, as demonstrated by the reduced clot weight and % IE values. It should be noted that even though the vessel is reperfused, some residual clot strands remain.

VASCULAR GRAFTS

The use of prosthetic grafts in small-diameter arterial reconstruction is necessary when suitable autologous vessel is not available (Roohk et al., 1976). A major limitation, however, is the low patency rate of these grafts early postoperatively when graft fibers protrude from the luminal surface (Goldman et al., 1982). Since PTFE (the most commonly used graft biomaterial) has protruding fibers that are a similar diameter as collagen fibrils (Berger et al., 1974), (platelets are known to adhere to exposed endothelium, which contains collagen fibrils) it is not surprising that they accumulate on these materials as part of the hemostatic process (Forbes and Prentice, 1978; Hanson et al., 1980). Platelets are believed to have an initiating role in the formation of occlusive arterial thrombi, and therefore platelet inhibitors in the presence of various biomaterial grafts have been used, but these results may not accurately reflect the platelet function *in vivo* (Callow et al., 1982; Moore et al., 1981). ^{111}In-labeled platelets have been used to evaluate *in vivo* abnormal platelet deposition on arterial grafts (Ritchie et al., 1981; Fuster et al., 1979; Dewanjee et al., 1978; Christenson et al., 1981; Radia et al., 1981). This imaging technique may also be useful in the elucidation of the platelet-to-vessel interaction, the production of pseudointima (Hagen et al., 1982), and the mechanism of platelet inhibitory drugs on vascular grafts (Clagett et al., 1982).

This technique of dual isotope subtraction was applied to a canine model of prosthetic Dacron graft implants. The canine is an accelerated model for the human reaction so that measurements at about 3 months in the dog are equivalent to 2 yr in humans. The canine model employed readily lends itself to further studies to evaluate antithrombotic therapy to reduce platelet deposition and increase prosthetic vascular graft patency. *In vivo* examination of platelet deposition as determined by ^{111}In-labeled platelets will define, it is hoped, the period during which antiplatelet therapy would be clinically advantageous (Rao, 1981). Using a standard surgical technique, 4-mm ID vascular grafts 5 cm in length were placed end to end in the carotid and femoral arteries in a series of 11 mongrel dogs. Each dog had four different grafts, autologous jugular vein, PTFE, and two knitted Dacron velour grafts.

Radiolabeled autologous platelets and RBC were administered and blood flow to the grafts was restored. Images were obtained immediately after surgery, and 1, 7, 14, 21, 30, 60 and 90 days after surgery. The regions corresponding to the grafts were identified; counts corresponding to the regions were obtained so that the % IE could be calculated. The mean % IE for each graft material was compared. The difference between the vein and the two Dacron velour grafts is statistically significant ($p < .01$ and $p < .05$, respectively), but the difference between Dacron and vein is insignificant. Analysis of the calculated platelet deposition (represented as % IE) revealed a significant linear relationship between the *in vivo* and *in vitro* measurement. These results suggest that the % IE value obtained scintigraphically using ^{111}In-platelets will enable temporal experimental and clinical studies to be evaluated quantitatively.

In conclusion, it is important to choose the proper animal model to obtain directly or indirectly correlatable results in all of these models. The dual isotope subtraction technique enables one to quantitate scintigrams and platelet interaction in various phases of atherosclerosis, thrombosis, and drug intervention studies.

ACKNOWLEDGMENTS

The authors wish to thank their many colleagues who have been involved in the various aspects of this work. The authors are supported by NIH Grant NS 06833 and HL 14147.

REFERENCES

Aas, K. A., and Gardner, F. H. (1958). Survival of blood platelets labelled to chromium-51. *J. Clin. Invest.* **37:**1257–1268.

Adelman, B., Stemmerman, M. B., Mennell, D., and Handin, R. I. (1981). The interaction of platelets with aortic subendothelium: inhibition of adhesion and secretion by Prostaglandin I_2. *Blood* **58:**198–205.

Addonizio, V. P., Edmunds, L. H., and Colman, R. W. (1978). The function of monkey (*M. mulatta*) platelets compared to platelets of pig, sheep, and man. *J. Lab. Clin. Med.* **91:**989–997.

Anitschkow, N., and Chalatow, S. (1913). Uever excperimentelle cholesterinsteatose und inre bedeutung fur die entstehung einiger pathologischer prozesse. *Zentrabl. Allg. Pathol.* **24:**1–9.

Armstrong, J. M., Lattimer, N., Moncada, S., and Vane, S. R. (1978). Comparison of the vasodepressor effects of prostacyclin and 6-oxo-prostaglandin F_{1a} with those of prostaglandin E_2 in rats and rabbits. *Br. J. Pharmacol.* **62:**125–130.

Ashford, T. P., and Freiman, D. G. (1968). Platelet aggregation at sites of minimal endothelial injury. An electron microscopic study. *Am. J. Pathol.* **53:**599.

Aster, R. H. (1971). Factors affecting the kinetics of isotopically labeled platelets. In *Platelet Kinetics,* Paulus, J. M., ed. Amsterdam: North Holland, p. 5.

Astrup, T. (1969). Relation between fibrin formation and fibrinolysis. In *Thrombosis.* Sherry, S., Brinkhous, K. M., Genton, E., and Stingle, J. M. eds. Washington, D.C.: National Academy of Sciences, pp. 609–635.

Barlow, G. H. (1979). Pharmacology of fibrinolytic agents. *Prog. Cardiovasc. Dis.* **21:**315–326.

Barnhart, M. I., and Chen, S-T. (1978). Vessel wall models for studying interaction capabilities with blood platelets. *Semin. Thromb. Hemostasis* **5:**112–155.

Barndt, R., Blankenhorn, D. H., Crawford, D. W., and Brooks, S. H. (1977). Regression and progression of early femoral atherosclerosis in treated hyperlipoproteinemic patients. *Ann. Intern. Med.* **86:**139–146.

Baumgartner, H. R., and Spaet, T. H. (1970). Endothelial replacement in rabbit arteries. *Fed. Proc.* **29:**710.

Baumgartner, H. R., and Haudenschild, C. (1972). Adhesion of platelets to subendothelium. *Ann. N.Y. Acad. Sci.* **201:**22–36.

Baumgartner, H. R., Muggli, R., Turitto, V. T., and Tschopp, T. B. (1976). Platelet adhesion, release and aggregation in flowing blood: Effects of surface properties and platelet function. *Thromb. Haemostasis* **35:**124–138.

Baumgartner, H. R. (1977a). Platelet interaction with collagen fibrils in flowing blood. I. Reaction of human platelets with alpha chymotrypsin-digested subendothelium. *Thromb. Haemostasis* **37:**1–16.

Baumgartner, H. R., Tschopp, T. B., and Weiss, H. J. (1977b). Platelet interaction with collagen fibrils in flowing blood. II. Impaired adhesion-aggregation in bleeding disorders—A comparison with subendothelium. *Thromb. Haemostasis* **37:**17–28.

Berger, S., Salzman, E. W., Merrill, E. W., and Wong, P. S. (1974). The reaction of platelets with prosthetic surfaces. In *Platelets: Production, Function, Transfusion, and Storage*. Baldini, M. G., and Ebbe, S., eds., New York: Grune and Stratton, pp. 299–312.

Bergman, G., Atkinson, L., Richardson, P. J., Daly, K., Rothman, M., Jackson, G., and Jewitt, D. E. (1981). Prostacyclin: Haemodynamic and metabolic effects in patients with coronary artery disease. *Lancet* **1:**569–572.

Bergmann, S. R., Lerch, R. A., Fox, K. A. A., Ludbrook, P. A., Welch, M. J., Ter-Pogossian, M. M., and Sobel, B. E. (1982a). The temporal dependence of efficacy of intracoronary thrombolysis assessed by positron tomography. *Am. J. Med.* **73:**573–581.

Bergmann, S. R., Lerch, R. A., Mathias, C. J., Sobel, B. E., and Welch, M. J. (1982b). Non-invasive detection of coronary thrombi with ^{111}In-platelets. *J. Nucl. Med.* (in press).

Bergmann, S. R., Mathias, C. J., Sobel, B. E., and Welch, M. J. (1982c). Evaluation of thrombolytic therapy in coronary artery thrombosis: Scintigraphic detection with the use of In-111 labeled platelets. In *Nuclear Medicine and Biology,* Vol. I, Raynaud, C., ed. Paris: Pergamon Press, pp. 65–68.

Beswick, W., Chmiel, R., Booth, R., Villar, I., Gilford, E., and Chesterman, D. (1979). Detection of deep venous thrombosis by scanning of ^{99m}technetium-labelled red-cell venous pool. *Br. Med. J.* **1:**82–84.

Bjornson, J., and Aursner, I. (1977). The haemostatic effect of ^{51}Cr-labelled blood platelets. *Scand. J. Haematol.* **18:**326–332.

Blix, W. (1964). The stability of fibrinolytic inhibitors in human sera and the inhibitor content in animal sera. *Scand. J. Clin. Lab. Invest.* **16:**614–616.

Born, G. V. R., Honour, A. J., and Mitchell, J. R. A. (1964). Inhibition by adenosine and by 2-choroadenosine of the formation and embolization of platelet thrombi. *Nature* **202:**761–764.

Born, G. V. R. (1974). Aggregation of haemostatic cells as an example of specialized cell function. In *Platelet Aggregation and Drugs,* Caprino, L., and Rossi, E. C., eds. New York: Academic Press, pp. 1–19.

Born, G. V. R. chr. (1980). Proceedings of the British Institute of Radiology. Cell labeling with gamma emitting radionuclides for in vivo study. *Br. J. Radiol.* **53:**922–931.

Bourgain, R. H. (1979a). The inhibitory effect of PGI_2 (Prostacyclin) on white platelet arterial thrombus formation. *Haemostasis* **8:**117–119.

Bourgain, R. H., Andries, R., and Finne, E. (1979b). The role of prostaglandins on platelet–vessel wall interaction. *Arch. Int. Pharmacodyn Ther.* **239:**161–163.

Branemark, P. I., and Ekholm, R. (1968). Adherence of blood cells to vascular endothelium. A vital microscopic and electron microscopic investigation. *Blut* **16:**274.

Brinkhous, K. M. (1978a). Animal models: Importance in research on hemorrhage and thrombosis. In *Thrombosis. Animal and Clinical Models.* Day, D. H., Molony, B. A., Nishezawa, E. E., and Rynbrandt, R. H., eds. New York: Plenum Press, pp. 123–133.

Brinkhous, K. M., and Bowie, E. J. W. (1978). Animal models of atherosclerosis involving the thrombotic process. In *The Thrombotic Process in Atherogenesis,* Chandler, A. B., Eurenius, K., McMillan, G. C., Nelson, C. B., Schunrtz, C. J., and Wessler, S., eds. New York: Plenum Press, pp. 385–400.

Buchanan, M. R., Dejana, E., Gent, M., Mustard, J. F., and Hirsch, J. (1981). Enhanced platelet accumulation onto injured carotid arteries in rabbits after aspirin treatment. *J. Clin. Invest.* **67:**503–508.

Bullock, B. C., Lehner, N. D. M., Clarkson, T. B., Feldner, M. A., Wagner, W. D., and Lofland, H. B. (1975). Comparative primate atherosclerosis I. Tissue cholesterol concentration and pathologic anatomy. *Exp. Mol. Pathol.* **22**:151–175.

Callow, A. D., Connolly, R., O'Donnell, T. F., Gembarowicz, R., Kahn, P., Keough, E., Ramberg-Laskaris, K., Valeri, R., Richmand, D. M., and Farver, S. P. (1982). Platelet-arterial synthetic graft interaction and its modification. *Int. Soc. Cardiovasc. Surg.* June, pp. 28–29.

Clagett, G. P., Robinowitz, M., Maddox, Y., Langloss, J. M., and Ramwell, P. W. (1982). The antithrombotic nature of vascular prosthetic pseudointima. *Surgery* **91**:87–94.

Christenson, J. T., Megerman, J., Hanel, K. C., L'Italien, G. J., Strauss, H. W., and Abbott, W. M. (1981). The effect of blood flow rates on platelets deposition in PTFE arterial bypass grafts. *Trans. Am. Soc. Artif. Intern. Organs* **27**:188–191.

Clarkson, T. B. (1972). Animal models of atherosclerosis. *Adv. Vet. Sci. Comp. Med.* **16**:151–173.

Clarkson, T. B., Prichard, R. W., Netsky, M. G., and Lafland, H. B. (1959). Atherosclerosis in pigeons: Its spontaneous occurrence and resemblance to human atherosclerosis. *Arch. Pathol.* **68**:143–147.

Cohen, P., and McCombs, H. L. (1967). Platelets and atherogenesis I. Augmentation of cholestero atherogenesis in the rabbit by a phlebotomy programme designed to induce thrombocytosis. *Br. J. Exp. Pathol.* **48**:346–356.

Cliffton, E. E., and Downee, G. R. (1950). Variations in proteolytic activity of serum of animals including man. *Proc. Soc. Exp. Biol. Med.* **73**:559–562.

Coleman, R. E., Harwig, S. S., Harwig, J. F., Sherman, L. A., and Welch, M. J. (1975a). Radioiodinated soluble canine fibrin. *Circ. Res.* **37**:35–40.

Coleman, R. E., Harwig, S. S. L., Harwig, J. F., Siegel, B. A., and Welch, M. J. (1975b). Fibrinogen uptake by thrombi: Effect of thrombus age. *J. Nucl. Med.* **16**:370–373.

Coleman, R. E., Welch, D. M., Baker, W. J., and Beightol, R. W. (1980). Clinical experience using indium-111-labeled leukocytes. In *Indium-111 Labeled Neutrophils, Platelets, and Lymphocytes,* Thakur, M. L., and Gottschalk, A., eds. New York: Trivirum Pub. Co., pp. 103–118.

Constantine, J. W. (1965). Temperature and the aggregation of rabbit platelets by adenosine diphosphate. *Nature* **205**: 1075–1077.

Cotran, R. S. (1969). Ultrastructural studies of endothelial injury in the micro circulation with special reference to thrombosis. In *Thrombosis.* Sherry, S., Brinkhous, K. M., Genton, E., and Stengle, J. M., eds. *Natl. Acad. Sci.,* Washington, D.C., pp. 437–446.

Cowan, C. R., and Monkhouse, F. C. (1966). Studies on electrically induced thrombosis and related phenomena. Can. *J. Physiol. Pharmcol* **44**:881.

Cowley, M. J., and Gold, H. K. (1982). Use of intracoronary streptokinase in acute myocardial infarction. *Mod. Concepts Cardiovas. Dis.* **51**:97–102.

Creasy, D. M., Fallenfant, M., James, D. A., and Dayan, A. D. (1979). Preliminary toxicity testing of prostacyclin. In *Prostacyclin.* Vane, J. R., and Bergstrom, S., eds. New York: Raven Press, pp. 385–391.

Cunningham, D. A., and Siegel, B. A. (1982). Radiolabeled platelets. In *Nuclear Medicine Annual 1982,* Freeman, L. M., ed. New York: Raven Press, pp. 143–165.

Danpure, H. J., Osman, S., and Brady, F. (1982). The labelling of blood cells in plasma with ^{111}In-tropolonate. *Br. J. Radiol.* **55:**247–249.

Davis, H. H., Siegel, B. A., Joist, J. H., Heaton, W. A., Mathias, C. J., Sherman, L. A., and Welch, M. J. (1978). Scintigraphic detection of atherosclerotic lesions and venous thrombi in man with Indium-111-labelled autologous platelets. *Lancet* **1:**1185–1187.

Davis, H. H., and Welch, M. J. (1979). Radioisotopic detection of arterial thrombi. In *Venous and Arterial Thrombosis: Pathogenesis, Diagnosis, Prevention, and Therapy.* Joist, J. H., and Sherman, L. A., eds. New York: Grune and Stratton, pp. 295–303.

Davis, H. H., Siegel, B. A., Sherman, L. A., Heaton, W. A., Naidich, T. P., Joist, J. H., and Welch, M. J. (1980a). Scintigraphic detection of carotid atherosclerosis with In-111-labeled autologous platelets. *Circ* **61:**982–988.

Davis, H. H., Siegel, B. A., Sherman, L. A., Heaton, W. A., and Welch, M. J. (1980b). Scintigraphy with ^{111}In-labeled autologous platelets in venous thromboembolism. *Radiology* **136:**203–207.

Dejana, E., Cozenave, J-P., Groves, H. M., Kinlough-Rathbone, R. L., Richardson, M., Packham, M. A., and Mustard, J. F. (1980). The effect of aspirin inhibition of PGI_2 production on platelet adherence to normal and damaged rabbit aortea. *Thromb. Res.* **17:**453–464.

DePalma, R. G., Insull, W., Bellow, E. M., Roth, W. T., and Robinson, A. V. (1972). Animal models for the study of progression and regression of atherosclerosis. *Surgery* **72:**268–278.

Dewanjee, M. K., Fuster, V., Kaye, M. P., and Josa, M. (1978). Imaging platelet deposition with ^{111}In-labeled platelets in coronary artery bypass grafts in dogs. *Mayo Clin. Proc.* **53:**327–331.

Dewanjee, M. K., Rao, S. A., and Didisheim, P. (1981). Indium-111 Tropolone, a new high-affinity platelet label: Preparation and evaluation of labeling parameters. *J. Nucl. Med.* **22:**981–987.

DeWood, M. A., Spores, J., Notske, R., Mouser, L. T., Burroughs, K., Galden, M. S., and Lang, H. T. (1980). Prevalence of total coronary occlusion during the early hours of transmural myocardial infarction. *N. Engl. J. Med.* **303:**897–902.

Didisheim, P. (1976). Animal models useful in the study of thrombosis and antithrombotic agents. In *Progress in Hemostasis and Thrombosis,* Vol. 1. Spaet, T. H., ed. New York: Grune and Stratton, pp. 165–197.

Dodds, W. J. (1978). Platelet function in animals: Species specificities. In *Platelets: A Multidisciplinary Approach.* Gaetano, G., and Garattini, S., eds. New York: Raven Press, pp. 45–59.

Doolittle, R. F., Omcley, J. L., and Surgenor, D. M. (1962). Species differences in the interaction of thrombin and fibrinogen. *J. Biol. Chem.* **237:**3123.

Duckert, F. (1979). Thrombolytic therapy in myocardial infarction. *Prog. Cardiovasc. Dis.* **21:**342–350.

Duval, D. L., Didisheim, P., Titus, J. L., Spittell, J. A., and Owen, C. A. (1970). Experimental arterial thrombosis: Description of a method. *Mayo Clin. Proc.* **45:**388.

Erhardt, L. R., Lundman, T., and Millstedt, H. (1973). Incorporation of ^{125}I-labelled fibrinogen into coronary arterial thrombi on acute myocardial infarction in man. *Lancet* **1:**387–390.

Ezekowitz, M. D., Smith, E. O., Cox, A. C., and Taylor, F. B. (1981a). Failure of aspirin to prevent incorporation of In-111-labeled platelets into cardiac thrombi in man. *Lancet*:440–443.

Ezekowitz, M. D., Leonard, J. C., Smith, E. O., Allen, E. W., and Taylor, F. B. (1981b). Identification of left ventricular thrombi in man using Indium-111-labeled autologous platelets. Circ **63:**803–809.

Finklestein, S., Miller, A., Callahan, R. J., Fallon, J. T., Godley, F., Feldman, B. L., Hinton, R. C., Roberts, A. B., Strauss, H. W., and Lees, R. S. (1982). Imaging of acute arterial injury with ^{111}In-labeled platelets: A comparison with scanning electron micrographs. *Radiology* **145:**155–159.

FitzGerald, G. A., Friedman, L. A., Miyamori, I., O'Grady, J., and Lewis, P. J. (1979). A double blind placebo controlled crossover study of prostacyclin in man. *Life Sci.* **25:**665–672.

Fletcher, A. P., and Sherry, S. (1960). Thrombolytic (fibrinolytic) therapy for coronary heart disease. Circ **22:**619–626.

Forbes, C. D., and Prentice, C. R. (1978). Thrombus formation and artificial surfaces. *Br. Med. Bull.* **34:**201–207.

French, J. E., Macfarlane, R. G., and Sanders, A. G. (1964). The structure of haemostatic plugs and experimental thrombi in small arteries. *Br. J. Exp. Pathol.* **45:**467–474.

French, J. E. (1969). The fine structure of experimental thrombi. In *Thrombosis,* Sherry, S., Binkhous, K. M., Genton, E., and Stengle, J. M., eds. *Natl. Acad. Sci.* Washington, D.C., pp. 300–320.

French, A., Hussey, J. K., Smith, F. W., Dendy, P. P., Bennett, B., and Douglas, A. S. (1981). Diagnosis of deep vein thrombosis using autologous indium-111 labelled platelets. *Br. Med. J.* **282:**1020–1022.

Friedman, M., and Byers, S. O. (1961). Experimental thrombo-atherosclerosis. *J. Clin. Invest.* **40:**1139.

Fuster, V., Dewanjee, M. K., Kaye, M. P., Josa, M., Metke, M. P., and Chesebro, J. H. (1979). Noninvasive radioisotopic technique for detection of platelet deposition in coronary artery bypass grafts in dogs and its reduction with platelet inhibitors. Circ **60:**1508–1407.

Gallimore, M. J., Nulkar, M. V., and Shaw, J. T. B. (1965). A comparative study of the inhibitors of fibrinolysis in human, dog, and rabbit blood. *Thromb. Diath. Haemorrh.* (*Stuttg.*) **14:**145–158.

Goldman, M., Norcott, H. C., Hawker, R. J., Drolc, Z., and McCollum, C. N. (1982). Platelet accumulation on mature dacron grafts in man. *Br. J. Surg.* **69** (Suppl):S38–S40.

Goodwin, D. A., Bushberg, J. T., Doherty, B. U., Lipton, M. J., Conley, F. K., Diamenti, C. J., and Meares, C. F. (1978). Indium-111-labeled autologous platelets for location of vascular thrombi in humans. *J. Nucl. Med.* **19:**626–634.

Gottlieb, H., and Lalich, J. J. (1954). The occurrence of arteriosclerosis in the aorta of swine. *Am. J. Pathol.* **30:**851–855.

Grimley, R. P., Rafiqi, E., Hawker, R. J., and Drole, Z. (1981). Imaging of ^{111}In-labelled platelets—a new method for the diagnosis of deep vein thrombosis. *Br. J. Surg.* **68:**714–716.

Grossman, Z. D., Wistow, B. W., McAfee, J. G., Subramanian, G., Thomas, F. D., Henderson, R. W., Rohner, R. F., and Roskopf, M. L. (1978). Tc-99m-oxine and In-111-oxine labeled platelets: Part II: Localization of experimentally induced vascular lesion. *J. Nucl. Med.* **19:**488–491.

Groves, H. M., Kinlough-Rathbone, R. L., Richardson, M., Moore, S., and Mustard, J. F. (1979). Platelet interaction with damaged rabbit aorta. *Lab. Invest.* **40:**194–200.

Gryglewski, R. J., Korbut, R., and Ocetkiewicz, A. (1978a). De-aggregatory action of prostacyclin *in vivo* and its enhancement by theophylline. *Prostaglandins* **15:**637–644.

Gryglewski, R. J., Korbut, R., and Ocetkiewicz, A. (1978b). Reversal of platelet aggregation by prostacyclin. *Pharmacol. Res. Commun.* **10:**185–189.

Guest, M. M., Daly, B. M., Ware, A. G., and Seegers, W. H. (1948). A study of antifibrinolyisin activity in the plasma of various animal species. *J. Clin. Invest.* **27:**785–792.

Hagen, P., Wang, S., Mikat, E. M., and Hackel, D. B. (1982). Antiplatelet therapy reduces aortic intimal hyperplasia distal to small diameter vascular prosthesis (PTFE) in nonhuman primates. *Ann. Surg.* **195:**328–339.

Hamberg, M., Svensson, B., and Samuelson, B. (1975). Thromboxanes–New group of biologically active compounds derived from prostaglandin endoperoxides. *Proc. Natl. Acad. Sci.* **72:**2994–2998.

Hanson, S. R., Harker, L. A., Ratner, B. D., and Hoffman, A. S. (1980). In vivo evaluation of artificial surfaces with a nonhuman primate model of arterial thrombosis. *J. Lab. Clin. Med.* **95:**289–304.

Harker, L. A., Ross, R., and Glomset, J. A. (1978). The role of endothelial cell injury and platelet response in atherogenesis. *Thromb. Haemostasis* (*Stuttg.*) **39:**312–321.

Harker, L. A. (1979). Platelet and fibrinogen kinetic evaluation of thrombogenesis. *Scand. J. Haematol. Suppl.* **34:**31–51.

Harris, R. H., Nichols, R., Schmeling, J. W., and Ramwell, P. W. (1981). Thromboxane A_2 and the endoperoxides mediate canine platelet activation. *Thromb. Res.* **23:**521–532.

Harwig, J. F., Harwig, S. S., Eichling, J. O., Colemen, R. E., and Welch, M. J. (1976). ^{123}I-Labeled soluble fibrin: preparation and comparison with other thrombus imaging agents. *Int. J. Appl. Rad. Isot.* **28:**157–162.

Harwig, S. S. L., Harwig, J. F., Sherman, L. A., Coleman, R. E., and Welch, M. J. (1977). Radioiodinated plasminogen: An imaging agent for pre-existing thrombi. *J. Nucl. Med.* **18:**42–45.

Hauss, W. J., Wissler, R. W., and Lehmann, R. eds. (1978). State of prevention and therapy in human arteriosclerosis and in animal models, ABH der Rhein-westg Akademie der Wisenschaften 63. Opladen, West Germany, Verlag.

Hawkey, C. M. (1970). Fibrinolysis in animals. In *The Haemostatic Mechanism in Man and Other Animals,* Macfarlane, R. G., ed. London: Academic Press, pp. 143–150.

Hawkey, C. M. (1974). The relationship between blood coagulation and thrombosis and atherosclerosis in man, monkeys and carnivores. *Thromb. Diath. Haemorrh.* **31:**103–118.

Heaton, W. A., Dais, H. H., Welch, M. J., Mathias, C. J., Joist, J. H., Sherman, L. A., and Seigel, B. A. (1979). Indium-111: A new radionuclide label for studying human platelet kinetics. *Br. J. Haematol.* **42:**613–622.

Hedner, U., and Nilsson, I. M. (1981). The role of fibrinolysis. *Clin. Haematol.* **10:**327–342.

Hendershott, L., Gentilcore, R., Ordway, F., Fletcher, J., and Donati, R. (1982).

Tropolone: A lipid solubilizing agent for cationic metals. *Eur. J. Nucl. Med.* **7**:234–236.

Heyns, A., Lotter, M. G., Badenhorst, P. N., vanReenen, O. R., Pieters, H., Minnaar, P. C., and Retief, F. P. (1980). Kinetics, distribution, and sites of destruction of ^{111}Indium-labelled human platelets. *Br. J. Haematol.* **44**:269–280.

Higgs, E. A., Moncada, S., and Vane, J. R. (1978). Effect of prostacyclin (PGI_2) on platelet adhesion to rabbit arterial subendothelium. *Prostaglandins* **16**:17–22.

Hirsch, E., and Loewe, L. (1946). A method for producing experimental venous thrombosis. *Proc. Soc. Exp. Biol. Med.* **63**:569.

Holmsen, H., Day, H. J., and Stormorken, H. (1969). The blood platelet release reaction. *Scand. J. Haematol. Suppl.* **8**:3–26.

Holmsen, H. (1974). Are platelet shape change, aggregation, and release reaction tangible manifestations of one basic platelet function. In *Platelets: Production, Function, Transfusion, and Storage,* Baldini, M., and Ebbe, S., eds. New York: Grune and Stratton, pp. 207–220.

Honour, A. J., Carter, R. D., and Mann, J. I. (1978). The effects of changes in the diet on lipid levels and platelet thrombus formation in living blood vessels. *Br. J. Exp. Pathol.* **50**:390.

Hornstra, G., and Haddeman, E. (1977). Effects of dietary fats on the role of platelets in arterial thromboembolism. In Platelets and Thrombosis, Mills, D. C. B., and Pareti, I., eds. London: Academic Press, pp. 115–124.

Hughes, J. (1962). Accolement des plaquettes aux structures conjonetines perivascularies. *Thromb. Diath. Haemorrh.* **8**:241.

Hurley, P. J. (1974). Red cell and plasma volumes in normal adults. *J. Nucl. Med.* **16**:46–52.

Jaeger, S., and Berntsen, H. (1979). Deposition of human labeled platelets on damaged rabbit aorta before and after ingestion of acetylsalicyclic acid. *Haemsostasis* **8**:99–105.

Jamieson, G. A. (1974). Interaction of platelets and collagen. In *Platelets: Production, Function, Transfusion, and Storage*. Baldini, M. G., and Ebbe, S., eds. New York: Grune and Stratton, pp. 171–176.

Johnson, S. A. (1968). Formation of thrombi on injured endothelium in mesenteric arterioles in guinea pig. *Thromb. Diath. Haemorrh. Suppl.* **28**:65–82.

Karim, S. M. M., and Adarkan, P. G. (1979). Some pharmacological studies with prostacyclin in baboon and man. In *Prostacyclin,* Vane, J. R., and Bergstrom, S., eds. New York: Raven Press, pp. 419–434.

Karpatkin, S., Charmatz, A., and Langer, R. M. (1970). Glycogeneses and glyconeogeneses in human platelets. *J. Clin. Invest.* **49**:140.

Kefalides, N. A., and Wingler, R. J. (1966). The chemistry of glomerular basement membrane and its relation to collagen. *Biochemistry* **5**:702.

Kien, M., Belamarich, F. A., and Shepro, D. (1971). Effect of adenosine related compounds on thrombocyte and platelet aggregation. *Am. J. Physiol.* **220**:604–608.

Knight, L. C., Primeau, J. L., Siegel, B. A., and Welch, M. J. (1978). Comparison of In-111-labeled platelets and iodinated fibrinogen for the detection of deep vein thrombosis. *J. Nucl. Med.* **19**:891–894.

Kordenat, R., and Kizidi, K. (1972). Experimental intracoronary thrombosis and selective in situ lysis by catheter technique. *Am. J. Cardiol.* **30**:640.

Kottke, B. A., and Subbiah, M. T. R. (1978). Pathogenesis of atherosclerosis: Concepts based on animal models. *Mayo Clin. Proc.* **53**:35–48.

Kramsch, D. M., and Hollander, W. (1968). Occlusive atherosclerotic disease of the coronary arteries in monkey (*Macaca irus*) induced by diet. *Exp. Mol. Pathol.* **9:**1–22.

Krohn, K. A., and Welch, M. J. (1977). Radiopharmaceuticals for thrombus detection: Selection, preparation, and critical evaluation. *Sem. Nucl. Med.* **7:**219–228.

Lewis, G. P., and Westwick, J. (1977). An in vivo model for studying arterial thrombosis. In *Thromboembolism: A New Approach to Therapy,* Mitchell, J. R. A., and Domenet, J. G., eds. New York: Academic Press, pp. 40–54.

Lewis, J. C., Fuster, V., and Kottke, B. A. (1977). Spontaneous endothelial cell injury in the intimal cushions of atherosclerotic pigeons. *Prog. Biochem. Pharm.* **14:**220–224.

Lewis, S. M., ICSH secretary, Panel on Diagnostic Application of Radioisotopes in Hematology (1977). Recommended methods for radioisotope platelet survival studies. *Blood* **50:**1137–1144.

Lutz, B. R., Fulton, G. P., and Akens, R. P. (1951). White thromboembolism in the hamster cheek pouch after trauma, infusion and neoplasia. Circ **3:**339.

MacFarlane, R. G. (1972). Haemostasis. In *Human Blood Coagulation, Haemostasis and Thrombosis,* Briggs, R., ed. London: Blackwell, pp. 543–585.

MacMillan, D. C., and Sim, A. K. (1970). A comparative study of platelet aggregation in man and laboratory animals. *Thromb. Diath. Haemorrh.* **24:**385–394.

McAfee, J. G., and Thakur, M. L. (1976). Survey of radioactive agents for in vitro labeling of phagocytic leukocytes. *J. Nucl. Med.* **17:**480–487.

McIlmoyle, G. Davis, H. H., Welch, M. J., Primeau, J. L., Sherman, L. A., and Siegel, B. A. (1977). Scintigraphic diagnosis of experimental pulmonary embolism with In-111-labeled platelets. *J. Nucl. Med.* **18:**910–914.

Majno, G., and Palade, G. E. (1961). Studies on inflammation. I. The effect of histamine and serotonin on vascular permeability. An electron microscopic study. *J. Biophys. Biochem. Cytol.* **11:**571.

Malinow, M. R. (1980). Atherosclerosis: Regression in nonhuman primates. *Circ. Res.* **46:**311–320.

Malmros, H., and Sternby, N. H., (1968). Induction of atherosclerosis in dogs by a thriouracil-free semisynthetic diet containing cholesterol and hydrogenated coconut oil. *Prog. Biochem. Pharm.* **4:**482–487.

Mason, R. G., and Read, M. S. (1971). Some species differences in fibrinolysis and blood coagulation. *J. Biomed. Mater. Res.* **5:**121–128.

Mathias, C. J., Heaton, W. A., Welch, M. J., Douglas, P. G., and Kelly, J. D. (1981). Comparison of ^{111}In-oxine and ^{111}In-acetylacetone for the labeling of cells: in vivo and in vitro biological testing. *Int. J. Appl. Rad. Isot.* **32:**651–656.

Mehta, P., and Mehta, J. (1979). Platelet function studies in coronary artery disease. V. Evidence for enhanced platelet microthrombus formation activity in acute myocardial infarction. *Am. J. Cardiol.* **43:**757–760.

Mehta, P., and Mehta, J. (1981). Role of platelet and endothelium in vascular disease. In *Platelets and Prostaglandins in Cardiovascular Disease,* Methta, J., and Mehta, P., eds. New York: Futura, pp. 1–21.

Mills, D. C. B. (1970). Platelet aggregation and platelet nucleotide concentration in various species. In *Haemostasis in Man and Other Animals,* Macfarlane, R. G., ed. London: Academic Press, pp. 99–107.

Moncada, S., Gryzlewski, R., Bunting, S., and Vane, J. R. (1976). An enzyme isolated from arteries transforms prostaglandin endoperoxides to an unstable substance that inhibits platelet aggregation. *Nature* **263:**663–665.

Moncada, S., and Vane, J. R. (1978). Unstable metabolites of arachidonic acid and their role in haemostasis and thrombosis. *Br. Med. Bull.* **34:**129–135.

Moncada, S., and Vane, J. R. (1979). Arachidonic acid metabolites and the interactions between platelets and blood vessel walls. *N. Engl. J. Med.* **300:**1142–1147.

Moore, S., and Inatocoycz, I. O. (1977). Vessel injury and atherosclerosis. *Adv. Exp. Med. Biol.* **102:**145–163.

Moore, W. S., Chvapil, M., Seiffert, G., and Keown, K. (1981). Development of an infection-resistant vascular prosthesis. *Surgery* **116:**1403–1407.

Moschos, C. B., Oldewurtel, H. A., Lahiri, K., Monshopf, G., and Regan, T. J. (1974). Incorporation of ^{131}I-fibrinogen in a coronary artery thrombus, detected in vivo with a scintillation camera. *Cardiovasc. Res.* **8:**715–720.

Moser, K. M., Grusan, M., and Bartimo, E. E. (1973). In vivo and post mortem dissolution rates of pulmonary emboli and venous thrombi in the dog. Circ **48:**170–178.

Moser, K. M., Spragg, R. G., Bender, F., Konopha, R., Hartman, M. T., and Fedullo, P. (1980). Study of factors that may condition scintigraphic detection of venous thrombi and pulmonary emboli with indium-111-labeled platelets. *J. Nucl. Med.* **21:**1051–1058.

Mustard, J. F., and Packham, M. A. (1970). Factors influencing platelet function: Adhesion, release, and aggregation. *Pharm. Rev.* **22:**97–187.

Mustard, J. F. (1976). Function of blood platelets and their role in thrombosis. *Trans. Am. Clin. Climatol. Assoc.* **87:**104–127.

Mustard, J. F. (1977a). Atherosclerosis, thrombosis, and clinical complications. In *Thromboembolism: A New Approach to Therapy,* Mitchell, J. R. A., and Domenet, J. G., eds. New York: Academic Press, pp. 3–25.

Mustard, J. F., Moore, S., Packham, M. A., and Kinlough-Rathbone, R. L. (1977b). Platelets, thrombosis and atherosclerosis. *Prog. Biochem. Pharmacol.* **14:**312–325.

Mustard, J. F., Packham, M. A., and Kinlough-Rathbone, R. L. (1977c). Platelets and its complications. In *Thrombosis: Animal and Clinical Model,* Day, H. J., Molony, B. A., Nushizawn, E. E., Rynbrandt, R. H., eds. New York: Plenum Press, p. 14.

Mysliwiec, M., Villa, S., Kornblihtt, T., Gaitano, G., and Donati, M. B. (1980). Decreased plasminogen activator but normal prostacyclin activity in rat veins during development of experimental thrombosis. *Thromb. Res.* **18:**159–165.

Nalbandian, R. M., and Henry, R. L. (1978). Platelet–endothelial cell interactions. *Sem. Thromb. Haemostasis* **5:**87–111.

Nam, S. C., Lee, W. M., Garmalyck, J., Lee, K. T., and Thomas, W. A. (1973). Rapid production of advanced atherosclerosis in swine by a combination of endothelial injury and cholesterol feeding. *Exp. Med. Pathol.* **18:**369–379.

Niewiarowski, S., and Latallo, Z. (1959). Comparative studies of the fibrinolytic system of sera of various vertibrates. *Thromb. Diath. Haemostasis* **3:**404–417.

Niewiarowski, S., Regoeczi, E., Stewart, G. J., Senyi, A., and Mustard, J. F. (1972). Platelet interaction with polymerizing fibrin. *J. Clin. Invest.* **51:**685–700.

Niewiarowski, S., and Steward, C. J. (1978). Interaction of blood cells with fibrinogen and polymerizing fibrin. In *Platelets: A Multidisciplinary Approach,* Gaetano, G., and Garatlini, S., eds. New York: Raven Press, pp. 131–147.

Ogston, D., and Bennett, B. (1978). Surface-mediated reactions in the formation of thrombin, plasmin and kallikrein. *Br. Med. Bull.* **34:**107–112.

Packham, M. A., Neshizawa, E. E., and Mustard, J. F. (1968). Response of platelets to tissue injury. *Biochem. Pharm.* (Suppl.) 171–184.

Packham, M. A., and Mustard, J. F. (1977). Clinical pharmacology of platelets. *Blood* **50:**555–573.

Pastakia, B. (1978). [111]Indium-labeled platelets in the management of idiopathic thrombocytopenic purpura. *Mayo Clin. Proc.* **53:**686.

Pearson, T. A., Kramer, E. C., Salz, K., and Hystinstall, K. H. (1977). The human atherosclerotic plaque. *Am. J. Pathol.* **86:**657–664.

Persson, B. R., and Kemper, V. (1977). Labeling and testing of ^{99m}Tc-streptokinase for the diagnosis of deep vein thrombosis. *J. Nucl. Med.* **16:**474–477.

Pick, R., Johnson, P. J., and Glick, G. (1974). Deleterious effects of hypertension on the development of aortic and coronary atherosclerosis in stumptail macaques (*Macaca speciosa*) on an atherogenic diet. *Circ. Res.* **35:**472–482.

Pick, R., Chediak, J., and Glick, G. (1978). Aspirin inhibits development of coronary atherosclerosis in cynomolgus monkeys (*Macaca fascicularis*) fed an atherogenic diet. *J. Clin. Invest.* **63:**158–162.

Powers, W. J., Mathias, C. J., Welch, M. J., Sherman, L. A., Siegel, B. A., and Clarkson, T. B. (1982a). Scintigraphic detection of platelet deposition in atherosclerotic macaques: A new model for investigation of antithrombotic drugs. *Thromb. Res.* **25:**137–142.

Powers, W. J., Seigel, B. A., Davis, H. H., Mathias, C. J., Clark, H. B., and Welch, M. J. (1982b). Evaluation of cerebrovascular disease with indium-111 platelet scintigraphy. *Neurology* **32:**938–943.

Powers, W. J., Mathias, C. J., Hopkins, K. T., Siegel, B. A., and Welch, M. J. (1982c). Dual radiotracer technique for improved scintigraphic detection of thrombi. In *Nuclear Medicine and Biology*, Vol. 1, Raynaud, C., ed. Paris: Pergamon Press, pp. 1163–1167.

Radia, R., Peters, A. M., Dunmamode, M., Fitzpatrick, M. L., Warrick, L. N., Huang, T. W., and Harker, L. A. (1981). In-111 platelet imaging for detection of platelet deposition in abdominal aneurysms and prosthetic arterial grafts. *Am. J. Cardiol.* **47:**882–889.

Ramirez, C. A., Stemerman, M. B., Isaacson, K. A., Colton, C. K., Smith, K. A., and Lees, R. S. (1981). Morphological and morphometric characterization of platelet adhesion to the exposed subendothelium of the rabbit thoracic aorta in vivo. *Microvasc. Res.* **21:**320–331.

Rao, G. H. R., Johnson, G. J., Reddy, R. K., and White, J. G. (1981). Rapid return of cyclo-oxygenase active platelets in dogs after a single oral dose of aspirin. *Prostaglandins* **22:**761–772.

Rentrop, P., Blanke, H., Karsch, K. R., Kaiser, H., Kostering, H., and Leitz, K. (1981). Selective intracoronary thrombolysis in acute myocardial infarction and unstable angina pectoris. Circ **63:**307.

Rhodes, B. A., Bill, W. R., Malmud, L. S., Siegel, M. E., and Wagner, H. N. (1973). Labelling and testing of urokinase and streptokinase: New Tracers for the detection of thromboemboli. In *Radiopharmaceuticals and Labelled Compounds,* vol. 2. Vienna, IAEA, pp. 163–170.

Riba, A. L., Thakur, M. L., Gottschalk, A., and Zaret, B. L. (1979). Imaging experimental coronary artery thrombosis with In-111 platelets. *Circulation* **60:**767–775.

Ritchie, J. L., Stratton, J. R., Thiele, B., Hamilton, G. W., Warrick, L. N., Huang, T. W., and Harker, L. A. (1981). In-111 platelet imaging for detection of platelet

deposition in abdominal aneurysms and prosthetic arterial grafts. *Am. J. Cardiol.* **47:**882–889.
Roohk, H. V., Pick, J., Hill, R., Hung, E., and Bartlett, R. H. (1976). Kinetics of fibrinogen and platelet adherence to biomaterials. *Trans. Am. Soc. Artif. Intern. Organs* **22:**1–7.
Roper-Drewinko, P. R., Drewinko, B., Corrigni, G., Johnston, D., McCredie, K. B., and Freireuk, E. J. (1981). Standardization of platelet function test. *Am. J. Hematol.* **11:**767–775.
Ross, R., and Bornstein, P. (1969). The elastic fiber I. The separation and partial characterization of its macromolecular components. *J. Cell. Biol.* **40:**366.
Ross, R., and Harker, L. (1976). Hyperlipidemia and atherosclerosis. Chronic hyperlipidemia initiates and maintains lesions by endothelial cell desquamation and lipid accumulation. *Science* **193:**1094–1100.
Ross, R. (1979). The arterial wall and atherosclerosis. *Ann. Rev. Med.* **30:**1–15.
Roth, G. J., Stanford, N., Jacobs, J. W., and Majerus, P. W. (1975). Acetylation of prostaglandin synthetase by aspirin purification and properties of acetylated proteins and sheep vesicular gland. *Proc. Natl. Acad. Sci.* **72:**3073–3076.
Russell, H. E., Bradhom, R. R., and Lee, W. H. (1966). An evaluation of infusion therapy (including Dextran) for venous thrombosis. Circ **33:**839.
Salazar, A. E. (1961). Experimental myocardial infarction. *Circ. Res.* **9:**1351–1356.
Sawyer, P. N., and Pate, J. W. (1953). Bio-electric phenomena as an etiological factor in intravascular thrombosis. *Am. J. Physiol.* **175:**103–107.
Schafer, A. I., and Handen, R. I. (1979). The role of platelets in thrombotic and vascular disease. *Prog. Cardiovasc. Dis.* **22:**31–52.
Scheffel, U., McIntyre, P. A., Evatt, B., Dvornisky, J. A., Natarajan, T. K., Bolling, D. R., and Murphy, E. A. (1977). Evaluation of Indium-111 as a new high photon yield gamma-emitting "physiological" platelet label. *Johns Hopkins Med. J.* **140:**285–293.
Scheffel, U., Tsan, M., and McIntyre, P. A. (1979). Labeling of human platelets with (^{111}In)8-hydroxyquinoline. *J. Nucl. Med.* **20:**524–531.
Schmidt, K. G., Rasmussen, J. W., Arendrup, H., and Rasmussen, A. D. (1982). Comparative studies of the in vivo kinetics of indium-111 labeled and chromium-51 labeled human platelets. In *Nuclear Medicine and Biology,* Vol. I, Raynaud, C. ed. Paris: Pergamon Press, pp. 937–939.
Schwartz, K. D., and Krugee, M. (1977). Improvement in labeling erythrocytes with ^{99m}Tc-pertechnitate. *J. Nucl. Med.* **12:**323–324.
Schwartz, S. I. (1962). Effects of electric environment on thrombosis. *Clin. Neurosurg.* **10:**291–297.
Scrutton, M. C., and Egan, C. M. (1979). Devalent cation requirements for aggregation of human blood platelets and the role of the anti-coagulant. *Thromb. Res.* **14:**713–727.
Sheppard, B. C., and French, J. E. (1971). Platelet adhesion in the rabbit abdominal aorta following the removal of the endothelium. A scanning and transmission electron microscopical study. *Proc. Roy. Soc. Lond. (Biol.)* **176:**427.
Sherry, S. (1977). The role of the platelet in thrombosis. In *Platelets and Thrombosis,* Mills, D. C. B., and Pareti, F. I. eds. New York: Academic Press, pp. 111–114.
Sinakos, Z., and Caen, J. P. (1967). Platelet aggregation in mammalians (human, rat,

rabbit, guinea-pig, horse, dog) a comparative study. *Thromb. Diath. Haemorrh.* **17:**99–111.

Sinn, H., and Silvester, D. J. (1979). Simplified cell label with indium-111 acetylacetone. *Br. J. Radiol.* **52:**758–759.

Sinzinger, H., Clopath, P., and Silberbauer, K. (1980). The effect of ballooning on minipeg aortic prostacyclin formation—a time course. *Artery* **1:**23–29.

Small, D. M. (1977). Cellular mechanisms for lipid deposition in atherosclerosis. *N. Engl. J. Med.* **297:**873–929.

Smith, J. B., and Willis, A. L. (1971). Formation and release of prostaglandins by platelets in response to thrombin. *Br. J. Pharm.* **40:**545.

Smith, J. B., Sedar, A. W., Ingerman, C. M., and Silver, M. J. (1977). Prostaglandin endoperoxides: Platelet shape change, aggregation, and the release reaction. In *Platelets and Thrombosis*. Mills, D. C. B., and Pareti, F. I., eds. New York: Academic Press, pp. 83–95.

Spaet, T. H. (1977). Delayed consequences of endothelial removal from rabbit aortae. *Adv. Exp. Med. Biol.* **102:**165–173.

Stary, H. C., and Malinow, M. R. (1982). Ultrastructure of experimental coronary artery atherosclerosis in cynomologous macaques. *Atherosclerosis* **43:**151–175.

Steinberg, D. (1979). Research related to underlying mechanisms in atherosclerosis. Circ **60:**1559–1565.

Stemerman, M. B., and Ross, R. (1972). Experimental arterioscolerosis. I. Fibrous plaque formation in primates, an electron microscope study. *J. Exp. Med.* **136:**769–789.

Stratton, J. R., Ritchie, J. L., Hamilton, G. W., Hammermeister, K. E., and Harker, L. A. (1981). Left ventricular thrombi: in vivo detection by indium-111 platelet imaging and two dimensional echocardiography. *Am. J. Cardiol.* **47:**874–881.

Sutherland, G. B., King, M. E., Peerless, S. J., Vegina, W. C., Brown, G. W., and Chamberlain, M. J. (1982). Platelet interaction within giant intracranial aneurysms. *J. Neurosurg.* **56:**53–61.

Szczeklik, A., Gryglewski, R. J., Nizankowski, R., Musial, J. Pieton, R., and Mruk, J. (1978). Circulatory and anti-platelet effects of intravenous prostacyclin in healthy men. *Pharm. Res. Commun.* **10:**545–556.

Szczeklik, J., Szczeklik, A., and Nizankowski, R. (1980). Haemodynamic changes induced by prostacyclin in man. *Br. Heart. J.* **44:**254–258.

Takeda, Y., and Nakabayashi, M. (1974). Physicochemical and biological properties of human and canine plasmins. *J. Clin. Invest.* **53:**154–162.

Tang, S. S., and Frozinovic, M. M. (1977). The effects of pCO_2 and pH on platelet shape, change and aggregation for human and rabbit platelet-rich plasma. *Thromb. Res.* **10:**135–145.

Thakur, M. L., Welch, M. J., Joist, H. J., and Coleman, R. E. (1976). Indium-111-labeled platelets: Studies on preparation and evaluation of in vitro and in vivo functions. *Thromb. Res.* **9:**345–357.

Thakur, M. L., Coleman, R. E., and Welch, M. J. (1977). Indium-111-labeled leukocytes for the localization of abcesses: Preparation, analysis, tissue distribution and comparison with gallium-67-citrate in dogs. *J. Lab. Clin. Med.* **89:**217–228.

Tsao, C. (1979). Vascular substances that modulate blood-to-vessel interactions. *Artery* **5:**246–261.

Vane, J. R., and Moncada, S. (1980). The anti-thrombotic effects of prostacyclin. *Acta Med. Scand.* (*Suppl.*) **642:**11–22.

Vermylen, J. (1978). Physiology of haemostasis. In *Platelets: A Multidisciplinary Approach*. Gaetano, G., and Garottini, S., eds. New York: Raven Press, pp. 3–16.

Verstraete, M. (1978). Biochemical and clinical aspects of thrombolysis. *Sem. Hematol.* **15:**35–54.

Wagner, W. D., and Clarkson, T. B. (1975). Comparative primate atherosclerosis. II. A biochemical study of lipids, calcium, and collagen in atherotic arteries. *Exp. Mol. Pathol.* **23:**96–121.

Walsh, P. N. (1977). Collagen-platelet interaction in coagulation, hemostases and thrombosis. In *Platelets and Thrombosis,* Mills, D. C. B., and Pareti, F. I., eds. New York: Academic Press, pp. 125–143.

Wautier, J. L., and Coen, J. P. (1979). Pharmacology of platelet suppressive agents. *Sem. Thromb. Hemostasis.* **5:**293–315.

Weiss, H. J. (1974). Pathophysiology and detection of clinically significant platelet dysfunction. In *Platelet: Production, Function, Transfusion and Storage,* Baldini, M. G., and Ebbe, S., eds. New York: Grune and Stratton, pp. 253–264.

Weiss, H. J. (1975). Platelet physiology and abnormalities of platelet function. *N. Engl. J. Med.* **293:**531–541, 580–588.

Weiss, H. J., and Turitto, V. T. (1979). Prostacyclin (prostaglandin I_2, PGI_2) inhibits platelet adhesion and thrombus formation on subendothelium. *Blood* **53:**244–250.

Welch, M. J., and Krohn, K. A. (1975). Critical review of radiolabeled fibrinogen: Its preparation and use. In *Radiopharmaceuticals,* Subramanian, G., Rhodes, B. A., Cooper, J. F., and Sodd, V. J., eds. New York: Society of Nuclear Medicine. pp. 493–502.

Welch, M. J., and Mathias, C. J. (1980). Platelet viability following Indium-111-oxine labeling in electrolyte solutions. In *Indium-111-Labeled Neutrophils, Platelets, and Lymphocytes,* Thakur, M. L., and Gottschalk, A., eds. New York: Trivirum, pp. 93–100.

Welch, M. J., Mathias, C. J., and Siegel, B. A. (1980). Clinical experience with indium-111-labeled platelets. In *Indium-111-Labeled Neutrophils, Platelets, and Lymphocytes,* Thakur, M. L., and Gottschalk, A., eds. New York: Trivirum, pp. 171–176.

Welch, M. J., Mathias, C. J., Jacobs, D., Rubin, J., Siegel, B. A., and Needleman, P. (1981a). In vivo manipulation of platelet thrombi: Platelet reversal using prostacyclin. *Stroke* **12:**117.

Welch, M. J., Mathias, C. J., Powers, W. J., Sherman, L. A., Siegel, B. A., and Clarkson, T. B. (1981b). Use of In-111-labeled platelets to assess platelet deposition on arteriosclerotic plaques in a primate model. *J. Nucl. Med.* **22:**P55.

Welch, W. H. (1887). The structure of white thrombi. *Trans. Pathol. Soc. Phila.* **13:**281–300.

Welch, W. H. (1899). Diseases of blood-vessels: Thrombosis. In *A System of Medicine,* Vol. 6, Allbutt, T. C., ed. London: Macmillan, pp. 155–228.

Wennesland, R., Brown, E., and Hopper, J. (1959). Red cell, plasma, and blood volume in healthy men measured by radiochromium (Cr^{51}) cell tagging and hematocrit: influence of age, somatotype, and habits of physical activity on the variance after regression of volumes to height and weight combined. *J. Clin. Invest.* **38:**1065–1077.

White, A. M., and Heptinstall, S. (1978). Contribution of platelets to thrombus formation. *Br. Med. Bull.* **34:**123–128.

Williams, R. D., and Carey, B. S. (1959). Studies in the production of "standard" venous thrombosis. *Ann. Surg.* **149:**381.

Wissler, R. W., and Vesselinovitch, D. (1977). Atherosclerosis in nonhuman primates. *Adv. Vet. Sci. Comp. Med.* **21:**351–421.

Woolf, N. (1978). Thrombosis and atherosclerosis. *Br. Med. Bull.* **34:**137–142.

Wu, K. K., Armstrong, M. L., Hoak, J. C., and Megan, M. B. (1975). Platelet aggregates in hypersholesterolemic rhesus monkeys. *Thromb. Res.* **7:**917–924.

Wu, K. K. (1981). Studies of stroke by specific labeling of platelets and fibrinogen. In *Cerebrovascular Diseases*. Moosay, J., and Remmuth, O. M., eds. New York: Raven Press, pp. 307–312.

Zucker, M. B. (1980). The functioning of blood platelets. *Sci. Am.* **6:**86–103.

CHAPTER 6

Animal Models in the Study of Hepatobiliary Radiotracers

ALAN R. FRITZBERG and DUANE C. BLOEDOW

INTRODUCTION AND CHARACTERISTICS OF HEPATOBILIARY AGENTS

The last decade has been a time of active and fruitful interest in the development of hepatobiliary radiopharmaceuticals. In order to compare potential radiopharmaceuticals, define their biological behavior, and obtain data that would indicate behavior under clinical conditions, a variety of animal species and even isolated hepatocytes have been used. This chapter reviews the use of these animal models with respect to the type of information to be gained, advantages and disadvantages of the species under study, and correlation of data from one species to another as well as to the ultimate species, man.

Objectives for Clinical Use

Clinical use of hepatobiliary radiopharmaceuticals has defined several criteria that can be used to define the biological behavior of the ideal agent [Fritzberg and Klingensmith, 1982]. These criteria consist of rapid uptake of the radiopharmaceutical by the liver with an extraction efficiency approaching 100% and extraction by other pathways with an efficiency approaching 0%. In addition, the radioactivity should rapidly leave the hepatocyte and enter the bile caniliculi and then hepatic bile ducts. The resulting bolus of radioactivity allows visualization of the intra- and extrahepatic bile ducts. Beyond this point in the biliary tree, movement of radioactivity will be independent of the radiopharmaceutical as long as enterohepatic cycling is not significant. Because of these desired parameters, comparative evaluations of hepatobiliary agents will require animal studies that provide measurements of specificity as well as rates of uptake by the liver and excretion into the bile. Since patients often have reduced levels of hepatocellular function, it is also desirable to evaluate agents in animals under conditions that indicate their behavior when hepatocellular function is reduced.

Anatomy of Hepatobiliary Excretion

A diagrammatic representation (Fritzberg, 1981) of the hepatocyte and its relationship to blood supply and bile duct network is provided in Fig. 6-1. The smallest blood vessel of the liver is called the sinusoid. Lining the sinusoid are endothelial and Kupffer cells that belong to the reticuloendothelial system and phagocytize insoluble particles out of the blood. The flattened endothelial cells contain pores or fenestrae of various sizes that allow passage of plasma contents, excluding cells, into a perivascular space known as the space of Disse (Goresky, 1980). It is in this space that the compounds to be handled by the hepatocyte come into contact with the sinusoidal mem-

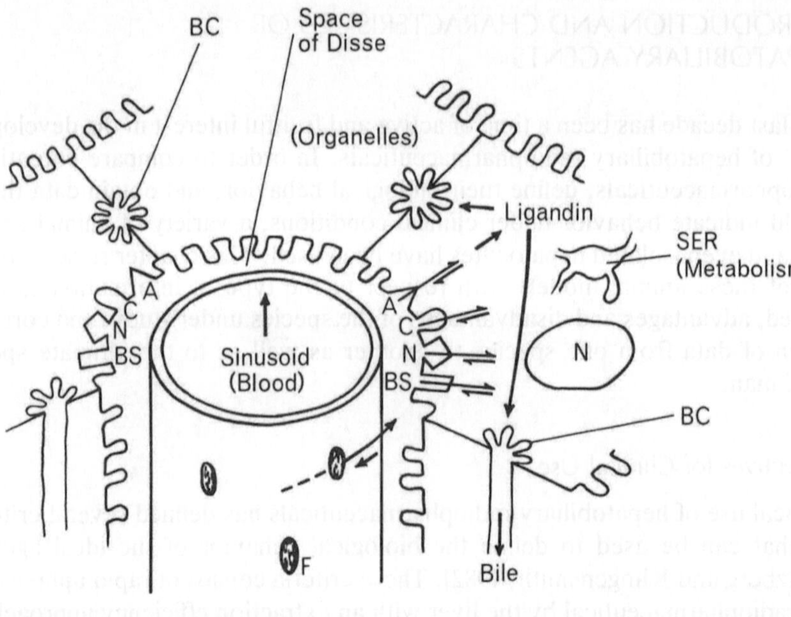

FIG. 6-1. Drawing of hepatocyte showing relationship of cellular parts to blood supply and biliary excretion. Materials in the blood diffuse through pores or fenestrae (F) in endothelial lining of sinusoid into the space of Disse. There they have access to sinusoidal membrane accepters for anions (A), cations (C), neutral compounds (N), or bile salts (BS). After transfer across the membrane they may be bound to intracellular proteins such as ligandin or metabolized at other sites such as the smooth endoplasmic reticulum (SER). Finally, transfer across the canilicular membrane results in passage into the bile via the bile caniliculus (BC) (Fritzberg, 1981).

brane surface of the hepatocyte. The sinusoidal membrane has evolved with many microvilli which is typical of membranes serving transport functions. After crossing the hepatocyte membrane the compounds may interact with cytoplasmic proteins or may be metabolized by hepatocyte organelles. Those compounds that are excreted into bile must then cross cell membranes that are common with other hepatocytes. A portion of these interhepatocyte membranes is specialized as canilicular or bile secreting and is referred to as the bile caniliculus. The bile caniculi are lined by plasma membrane of the hepatocyte with the microvilli. There is no direct communication between the space of Disse and the canilicular lumen. The canilicular lumen is sealed off by tight junctions known as close contact points or canulae occludentes (Reichen and Paumgartner, 1980). The canilicular space between hepatocytes leads to the ducts of Herring or preductules, ductules, and ducts as the biliary drainage network joins into larger bile-collecting vessels. The ductal network finally leads to the hepatic ducts, which in turn lead to the common bile duct, gallbladder, and other portions of the biliary tract. A secreting type of biliary epithelium is found at the level of ductules and ducts (Reichen and Paumgartner, 1980).

Biochemical Aspects of Hepatocyte Transport

Uptake of compounds by the hepatocyte and excretion into bile has been shown to be selective for a variety of classes of compounds. These classes include anions such as bilirubin and the dyes sulfbromophthalein (BSP), indocyanine green (ICG), and rose bengal (RB), bile acids, positively charged quarternary amines, and neutral steroid sugar conjugates (Arias, 1972). Specificity of hepatocyte uptake for these classes is indicated at the sinusoidal membrane (Fig. 6-1) by A for anions, BS for bile acid salts, N for neutral compounds, and C for cations. Uptake for these compounds is saturable and members within a class show competition for these membrane protein uptake sites.

Within the hepatocyte are cytosolic proteins that have been shown to bind bilirubin and anion dyes (Levi et al., 1969). One of these proteins, ligandin or Y protein (Litwack et al., 1971), which has been shown to be identical to glutathione transferase B (Habig et al., 1974), and Z protein form approximately 5% of the soluble protein of rat liver. It has been suggested (Arias et al., 1976), that ligandin influences hepatocyte uptake by controlling efflux of these compounds into plasma. Since ligandin also has glutathione transferase activity, it is able to conjugate glutathione to anions such as BSP, the form in which BSP is excreted into bile. For other types of metabolism before excretion, such as the glucuronidation of bilirubin, a further intracellular translocation to the smooth endoplasmic reticulum (SER) must take place.

The final excretory step involves transport across the canilicular membrane. Again, as with the sinusoidal membrane, specificity for classes of organic compounds is thought to exist (Schanker, 1968). These consist of at least three classes and include the dye anions, quarternary amine cations, and neutral steroid sugars.

Several mutant animal species have been useful in the study of compound class specificity for hepatobiliary excretion. For example, mutant Corriedale sheep have a depressed transport capacity for BSP, bilirubin, and other related anions but have normal secretory capacity for the bile acid taurocholate (Alpert et al., 1969). Hepatic uptake and storage of the dye anions are normal, however. This phenomenon has been used to substantiate different pathways for bilirubin and dye anions and for bile acids. In man this type of defect in hepatobiliary excretion results in increased levels of bilirubin and is referred to as Dubin–Johnson syndrome or congenital hyperbilirubinemia (Dubin and Johnson, 1954). Hereditary defects in which the recessive individual lacks bilirubin glucuronyl transferase are known. In man this deficiency is known as Crigler–Naijar syndrome, in which serum unconjugated bilirubin levels are in excess of 20 mg/dl (Arias et al., 1969). The Gunn rat also has this deficiency and has bilirubin levels in the range of 7–10 mg/dl. In addition, the Gunn rat also is unable to synthesize glucoside or xyloside conjugates of bilirubin (Fevery et al., 1972). Another related generic defect in man is Gilbert's syndrome, in which hepatic uptake of unconjugated bilirubin is impaired (Billing et al., 1964) and hepatic bilirubin glucuronyl transferase activity averages 20–50% of normal values (Black and Billing, 1969). A similar defect in animals, the mutant Southdown sheep, is also known (Mia et al., 1920). In such sheep the plasma clearance of dye anions is delayed and hepatic storage is reduced.

Hepatobiliary Excretion Pathway of ^{99m}Tc-Labeled Iminodiacetates

Two studies on ^{99m}Tc-labeled iminodiacetate hepatobiliary agents have been carried out in order to determine the mechanism by which these radiopharmaceuticals are excreted into bile. In dogs the hepatobiliary excretion of [^{99m}Tc]*N*-(2,6-dimethylacetanilide)iminodiacetate (Tc-HIDA, Tc-dimethyl-IDA) was depressed by BSP, but not by the quarternary amine oxyphenonium bromide (Harvey et al., 1979). In rats the excretion of [^{99m}Tc]*N*-(2,6-diethylacetanilide)iminodiacetate (Tc-diethyl-IDA) was depressed by BSP (Fritzberg et al., 1979) and indocyanine green (Fritzberg, 1981), but not by the quarternary amine procaine amide ethobromide or taurocholate (Fritzberg et al., 1979). These studies indicate that this class of hepatobiliary agent shares the pathway of bilirubin and other dye anions, but not bile acids or quarternary amine cations.

SPECIES AND HEPATOBILIARY EXCRETION

A number of hepatobiliary excretion parameters have been shown to be dependent on the species under study. These include fraction of the injected dose excreted via the hepatobiliary system, pharmacokinetic differences due to compartmental volumes and relative hepatic blood flow, anatomical differences, and bile flow differences. Any of these aspects are important to appreciate when data in one species are compared with those from another and when one is choosing an animal model.

Molecular Weight

The molecular weight of organic compounds has been shown to be a factor that determines the extent of hepatobiliary excretion, and it has also been found that species differences are significant with regard to this parameter.

In all species studied the biliary excretion of compounds, excluding quarternary ammonium compounds, of molecular weight (mol wt) less than 300 is low (5–10% of dose) (Smith, 1973). In the molecular weight range of 300–600 the excretion of compounds into bile can vary significantly. For example, 70–90% of the dose of stilbestrol glucuronide (mol wt 445) is excreted into bile in the rat, hen, dog, and cat, and 20–30% is excreted into bile in the guinea pig and rabbit (Smith, 1973). Phenolphthalein glucuronide (mol wt 495) resulted in 54% of the dose excreted in the bile of rats, 81% in dogs, 71% in hens, 34% in cats, 38% in sheep, 13% in rabbits, 6% in guinea pigs, and 9% in rhesus monkeys (Aziz et al., 1971).

Fritzberg et al. (1981) studied the biliary excretion of [^{99m}Tc]*N,N'*-bis-(mercaptoacetyl)ethylenediamine (DADS) (mol wt 319) (Jones et al., 1982) and found excretion at a low but significant level. In mice 7% (range 4.7–10.4%) of the dose was found in the bile. Comparable results were observed in rats (range 3.9–10.6%). The bile was not recovered in rabbits, but biliary excretion was evident in images obtained with the gamma camera. The total amount excreted in the urine was similar in mice, rats, and rabbits, suggesting that the biliary excretion in rabbits was also similar. In rats experiments in which the renal pedicles were tied off resulted in 19% biliary excretion in 90 min. Studies in renal transplant patients also showed hepatobiliary excretion as well as increased amounts in patients with decreased levels of renal function (Klingensmith et al., 1982). Thus the biliary pathway was clearly an alternative pathway for the complex. The consistent values for biliary excretion in all species are somewhat different from the trends in the literature described by molecular weight (Smith, 1973). That structural aspects are important was shown by studies of the carboxylate analog, [^{99m}Tc]*N,N'*-bis(mercaptoacetyl)-2,3-diaminopropanoate (mol wt 362), in which hepatobiliary excretion was 1% of the dose or less and increased to only 3% in 90 min in the absence of renal excretion in the rat (Fritzberg et al., 1982b).

Biliary excretion of organic anions of mol wt >600 is high in all species studies. Several generalizations have been made with regard to species, molecular weight, and hepatobiliary excretion. Threshold molecular weight values for hepatobiliary excretion of organic anions of 325 ± 50, 400 ± 50, and 475 ± 50 have been suggested for rats, guinea pigs, and rabbits, respectively (Hirom et al., 1967; Klaassen, 1975). Above 850 mol wt the excretion occurs mainly by the liver (Hirom et al., 1976). Based on their ability to excrete compounds into bile, species have been classified as good (rats, dogs, hens), intermediate (cats, sheep), or poor (rabbits, guinea pigs, rhesus monkeys) excretors (Klaassen, 1975; Smith, 1973).

Species differences have also been characterized for transport capacities of the liver for the dye anion BSP and bilirubin. For BSP transport maximum values of 1.13 (Klaassen and Plaa, 1967) and 1.3 μmol/min kg (Herz et al., 1973; Whelan and Combes, 1975) in the rat, 1.35 μmol/min kg in the rabbit (Klaassen and Plaa, 1967), 0.14 (O'Maille et al., 1966), and 0.20 μmol/min kg in dogs (Klaassen and Plaa, 1967), and 9.5 μmol/min in man (Dhumeaux et al., 1974). For bilirubin transport maxima of 1.04 μmol/min kg in rats (Weinbren and Billing, 1956), 0.51–0.73 μmol/min kg in mice (Roberts and Plaa, 1967; Roberts et al., 1968), 1.4–8.6 μmol/min kg in dogs, depending on bile flow (Goresky and Kleiger, 1969), 0.98 μmol/min kg in sheep (Upson et al., 1970) and 0.07 μmol/min kg in man (Raymond, 1966) have been observed. Man appears to stand out as having a lower capacity to excrete bilirubin than the animals commonly studied.

Significant differences in extents and rates of biliary excretion of radiopharmaceuticals in the range of 683–793 mol wt have also been observed. The biliary excretion of several radiotracers, including the dye anion [^{131}I]rose bengal, [^{99m}Tc]pyridoxylidene glutamate ([^{99m}Tc]PG), and [^{99m}Tc]*N*-(2,6-dimethylacetanilide)iminodiacetate ([^{99m}Tc]dimethyl-IDA), are compared in Table 6-1. Differences in amounts excreted in different species are apparent for each. The differences reflect mainly specificity of excretion, since all data were collected at 1 hr postinjection. Besides the difference in excretion due to species, the data also indicate structural and biliary excretion rate relationships. The molecular weights of the radiopharmaceuticals are similar, especially so for [^{99m}Tc]PG and [^{99m}Tc]dimethyl-IDA. In all species the biliary excretion differs significantly. That the trend is similar for each species is important for comparative evaluation of hepatobiliary agents.

Another way to consider species and radiotracer hepatobiliary excretion is to compare the amount excreted in the urine. Generally, retention in organs other than excretion via the hepatobiliary or renal pathways is insignificant. Table 6-2 shows such data for [^{99m}Tc]*N*-(2,6-dimethylacetanilide)iminodiacetate ([^{99m}Tc]dimethyl-IDA), [^{99m}Tc]*N*-(2,6-diethylacetanilide)iminodiacetate ([^{99m}Tc]diethyl-IDA), and [^{99m}Tc]pyridoxylideneglutamate ([^{99m}Tc]PG) in mice, rats, rabbits, dogs, baboons, and humans. It is of interest that the values for renal excretion in mice were relatively high when compared to the others. Rat values were generally low except for [^{99m}Tc]PG. Rabbit values were comparable to those for other species for [^{99m}Tc]dimethyl-IDA and [^{99m}Tc]diethyl-IDA but low for [^{99m}Tc]PG. Dog values were relatively high for all agents. Renal excretion in baboons and humans was

TABLE 6-1. Percent Injected Dose of Hepatobiliary Radiopharmaceuticals Excreted in Bile in One Hour in Different Species

Radiopharmaceutical (mol wt)	*Species*			
	Mouse	*Rat*	*Dog*	*Baboon*
[^{131}I]rose bengal (793)	67.2[a]	56.0[b]	20[c]	22[d]
^{99m}Tc-PG[e] (687)	42.5[f]	36.0[b]	—	17[d]
[^{99m}Tc]dimethyl-IDA (683)	68.3[g]	88.2[b]	55.2[c]	51[d]

[a] Loberg et al. (1979).
[b] Chervu et al. (1980).
[c] Harvey et al. (1979).
[d] Wistow et al. (1977).
[e] [^{99m}Tc]pyridoxylideneglutamate, molecular weight based on Kato-Azuma, 1981.
[f] Baker et al. (1975).
[g] Loberg et al. (1976).

TABLE 6-2. Renal Excretion of Hepatobiliary Radiopharmaceuticals in Different Species

Species	*[^{99m}Tc]dimethyl-IDA*[a]	*[^{99m}Tc]diethyl-IDA*[a]	*[^{99m}Tc]Tc-PG*[a]
Mice	35[b,c] (2 hr)	17[b] (2 hr)	47.8[d] (2 hr)
Rats	6.2[e] (1 hr)	4.2[f] (1 hr)	47.8[g] (1 hr)
Rabbits	15[h] (0.5 hr)	9.0[h] (0.5 hr)	18.9[h] (0.5 hr)
Dogs	13.3[i] (1 hr)	5.6[j] (2 hr)	56[k] (2 hr)
Baboons	20[l] (3 hr)	5.0[l] (3 hr)	28[l] (3 hr)
Humans	14.2[m] (1.5 hr)	6.8[n] (3 hr)	

[a] See text for complete chemical names.
[b] Values in parentheses are times postinjection.
[c] Chiotellis and Varavrigou (1980).
[d] Baker et al. (1975).
[e] Chervu et al. (1980).
[f] Data from ^{99m}Tc-SOLCO HIDA Product Information, SOLCO Ltd., Basel.
[g] Kato and Hazue (1978).
[h] Jansholt et al. (1979).
[i] Harvey et al. (1979).
[j] Fritzberg et al. (1982).
[k] Fritzberg, A. R., and Whitney, W. P., unpublished data.
[l] Wistow et al. (1977).
[m] Ryan et al. (1977).
[n] Klingensmith et al. (1981).

relatively high except for [^{99m}Tc]PG in baboons. In general, the range was not large for either [^{99m}Tc]dimethyl-IDA or [^{99m}Tc]diethyl-IDA except for the mice values. The values for [^{99m}Tc]PG did vary to a much greater extent. This may have been due to a greater variance in radiochemical purity for [^{99m}Tc]PG (Loberg et al., 1981). Importantly, in all cases the order was the same. That is, the renal excretion of [^{99m}Tc]PG was always greater than [^{99m}Tc]dimethyl-IDA, which was always greater than [^{99m}Tc]diethyl-IDA.

Anatomical Species Differences

With regard to anatomy the major species difference is the presence or absence of a gallbladder. The gallbladder is a pear-shaped organ connected to the common bile duct. The cystic duct controls flow out of the gallbladder and into the common bile duct. In most animals, it has a capacity of 1–2 ml/kg body weight. The principal function of the gallbladder is the concentration of sodium salts of bile acids in periods between eating. It absorbs sodium chloride, sodium bicarbonate, and water in isotonic proportions, and thus reduces the volume of hepatic bile 80–90% (Frizzell and Heintze, 1980). In some species, including the rat, deer, and whale, the gallbladder is absent. Since the gallbladder is part of the extrahepatic biliary network, the presence or absence of a gallbladder should not affect hepatic uptake and clearance of these agents.

Pharmacokinetic Species Differences

As discussed earlier, the optimal hepatobiliary radiopharmaceutical should have high specificity for liver uptake and biliary excretion and rapid kinetics such that the radiotracer moves into the liver, the intrahepatic ducts, and the lower biliary tract as a bolus. Crude comparisons can be made between agents by comparing biodistributions at early and late times postinjection and blood disappearance curves obtained by multiple blood sampling. However, significantly more information lies under the blood disappearance curve than is obtained by comparative rapidity of disappearance from the blood and uptake by the liver. Curve analysis and knowledge of parameters such as species blood flow to organs of interest allow the determination of pharmacokinetic parameters such as clearance, extraction efficiency, and volumes of distribution. These parameters provide a much better understanding of the *in vivo* behavior of the radiotracer. In addition, organ blood supply and body blood volumes for different species allow comparison of these parameters from studies in different species. Information may be gained as to whether or not species differences exist, and if they exist, what the basis is for the differences and whether extrapolations of parameter values to other species and to humans are reasonable.

Complex multicompartment pharmacokinetic models have been devised for the interpretation of the movement of hepatobiliary agents from the blood, through the liver, and into the bile (Smith et al., 1979; Gelius et al., 1981). For the initial evaluation of potential hepatobiliary agents and the determination of parameters that are useful in structure distribution relationships a simpler compartment-independent analysis may be used, however. The analysis begins with the time-dependent disappearances of the radiochemical from the plasma or blood following intravenous injection. Blood is preferred as a fluid of reference, since it is whole blood rather than plasma that perfuses the liver or organ of interest. In most cases, however, the interpretation of blood and plasma disappearance curves will be similar.

Derivation of Pharmacokinetic Parameters

The symbols used in the following discussion follow those suggested by Rowland and Tucker (1980). The disappearance of a radiotracer from the blood may often be described as a sum of exponentials of the form (Wilkinson and Shand, 1975):

$$C_b = \sum_{i=1}^{n} C_{b,i} e^{-\lambda_i t} \qquad \text{(Eq. 6-1)}$$

where C_b is the activity per volume in blood, $C_{b,i}$ and λ_i are constants of the exponential equation, n is the number of exponential terms required, and t is time after injection. For the simplest case, where $n = 1$, Eq. (6-1) reduces to Eq. (6-2). Reports in the literature describe

$$C_b = C_{b(0)} e^{-\kappa t} \qquad \text{(Eq. 6-2)}$$

blood disappearance of radiotracer using the term κ, which is a (first-order) rate constant for disappearance. The half-life of the disappearance of the radiotracer can be calculated from κ by the relationship

$$t_{1/2} = \frac{0.693}{\kappa} \qquad \text{(Eq. 6-3)}$$

While the rate constant (κ) and half-life ($t_{1/2}$) information are useful, both of these are hybrid parameters (Wilkinson and Shand, 1975) that depend on the distribution volume and the blood radiotracer clearance of the radiopharmaceutical according to Eq. (6-4):

$$\kappa = \frac{\text{blood clearance}}{\text{distribution volume}} \qquad \text{(Eq. 6-4)}$$

and

$$t_{1/2} = \frac{(0.693)(\text{distribution volume})}{\text{blood clearance}} \qquad \text{(Eq. 6-5)}$$

Although the interpretation of multiexponential [Eq. (6-1) with $n > 1$] blood radiopharmaceutical disappearance curves is more complex than the monoexponential class [Eq. (6-2)], the same general theory holds; that is, the disappearance half-life is directly proportional to distribution volume and inversely proportional to clearance. Thus, rather than using the half-life (or the rate constant for disappearance) when comparing radiotracers, one should look to the clearance as a measure of elimination efficiency. This type of analysis, rather than an examination of only the disappearance half-life, offers greater insight into the body's handling of the radiotracer.

Interpretation and Use of Parameters and Species Considerations
A *distribution volume* (V) for a radiopharmaceutical may be generally defined (Gibaldi and Perrier, 1975b) as a proportionality constant that relates the total radioactivity of the radiotracer within the body to the radioactivity of the radiotracer in the blood (C_b) or

$$V = \frac{\text{total radioactivity in body}}{C_b} \quad \text{(Eq. 6-6)}$$

For the monoexponential case [Eq. (6-2)], the concept is elementary, as only one distribution volume can be estimated using the intravenous injected dose in radioactivity units, and the radioactivity concentration within the blood at time zero $C_{b(0)}$, obtained by extrapolation, according to Eq. (6-7):

$$V = \frac{\text{dose}}{C_{b(0)}} \quad \text{(Eq. 6-7)}$$

The smallest distribution volume that is physiologically possible is the blood volume (about 5–7% of body weight in adult humans, or 5 liters). Distribution volumes of more than 5 liters in humans indicate that the radiotracer passes from the blood into extravascular spaces and tissues.

If the blood radiotracer disappearance curve is multiexponential [Eq. (6-1) with $N > 1$], the interpretation is more complex. Intuitively, however, the radiotracer must first distribute within the blood before it distributes into extravascular tissues. Thus the *initial* dilution volume (often referred to as the volume of the central compartment) may still be estimated using Eq. (6-7). Several other distribution volumes can be defined for the multiexponential blood disappearance curve (Gibaldi and Perrier, 1975a). One of these distribution volumes is referred to as the steady-state distribution volume (V_{ss}). This distribution volume is often considered the most appropriate as a true measure of distribution, since it is independent of the rate of elimination (Benet and Ronfeld, 1969). The steady-state distribution volume is easily estimated using only area under the blood disappearance curve by the method of Benet and Galeazzi (1979).

Two conditions must be met for the determination of V_{ss} from the coefficients and exponents of a polyexponential equation fit to blood concentration time data (Wagner, 1976). (1) All radiotracer transfers within the body depend on radiotracer concentration in a first-order fashion. (2) The radiopharmaceutical exits the body directly from the measured site. The first of these conditions is usually met, since the radiotracer is administered in high specific activity. In the case of rapidly cleared hepatobiliary agents, however, the second assumption may not be met. In the case of hepatobiliary agents this requires knowledge of whether the hepatocyte uptake or canilicular transport into bile is rate limiting. If the canilicular step is rate limiting, the elimination would be considered from a peripheral compartment, not the measured site, and V_{ss} as calculated may be underestimated.

Although there are both important and subtle differences in the interpretation of the various distribution volumes for the multiexponential case, it should simply be remembered that the magnitude of the distribution volume (when compared to blood volume for a given species) gives an indication of the extent to which the radiotracer passes from the blood into extravascular tissues.

Since the smallest distribution volume that is physiologically feasible is that of blood volume, the blood volume for each species under study should be known. Such blood volume information provides a true physiologic context in which to evaluate the distribution of the radiotracer. Blood (and plasma) volume information for a variety of species is available (Altman and Dittmer, 1972) and values for common laboratory animal models are given in Table 6-3. An alternative approach to actual blood volume measurements in a given animal species is that of the heterogonic equation relating blood weight (grams) to body weight (*B* in grams) (Adolph, 1949):

$$\text{Blood weight} = 0.055B^{0.99} \qquad \text{(Eq. 6-8)}$$

A radiopharmaceutical distribution volume of greater than blood volume in a given species indicates that the radiotracer can be found in extravascular tissues. Without biodistribution studies, the tissues into which the radiopharmaceutical is located cannot be readily identified. However, the ex-

TABLE 6-3. Blood Volumes and Hepatic Blood Flows in Various Species[a,b]

			Hepatic Blood Flow		
Species	*Body Weight (kg)*	*Blood Volume (ml)*	*ml/kg*	*flow ml/min*	*ml/min/kg*
Mouse	0.0304	1.62[c]	53.3	2.62	86.2
Rat	0.249	16.0	64.3	17.2	69.1
Rabbit	2.75	153	55.6	122	44.4
Dog	16.5	1420	86.1	676	41.0
Monkey	4.84	262	54.1	250	51.7
Human	70.0	4540	64.9	1780	25.4

[a] Altman and Dittmer (1972).
[b] Boxenbaum (1980).
[c] Calculated from relationship of body weight and blood weight.

travascular distribution of the radiopharmaceutical may be further interpreted using the relationship (Wilkinson and Shand, 1975; Gillette, 1971; Gibaldi, et al., 1978)

$$V = V_b + \frac{V_T fu_b}{f_T} \qquad \text{(Eq. 6-9)}$$

where V is the measured radiopharmaceutical distribution volume (V for monoexponential case, V_{ss} for multiexponential case), V_b is the actual blood volume, V_T is the volume of the tissue water (total body water volume minus blood volume), fu_b is the free (or unbound) fraction of radiotracer in the blood, and f_T is the free (or unbound) fraction of radiotracer in the extravascular tissues. This relationship describes the theoretical dependence of the radiotracer distribution volume and body fluid volumes (V_b and V_T) and a binding of the radiotracer within the blood (expressed as the bound fraction fu_b and within the tissues (expressed as the unbound fraction f_T). Given the distribution volume for a radiotracer in a certain species, the parameters in Eq. (6-9), with the exception of f_T, are either measurable or readily available. The fluid volumes are known based on reported blood and total body water volumes (Altman and Ditter, 1972a, 1972b; Adolph, 1949). The free fraction of radiotracer in blood may be measured by various techniques (Kurz et al., 1977). The only parameter of Eq. (6-9) that is not known therefore is the free fraction of radiotracer in the tissues (f_T). The f_T can be estimated by solving Eq. (6-9):

$$f_T = \frac{V_T \cdot fu_b}{V - V_b} \qquad \text{(Eq. 6-10)}$$

It is suggested that, in addition to estimating the radiotracer distribution volume and the free fraction of radiotracer in blood ($fu_b = 1 -$ bound fraction), the free fraction of radiopharmaceutical in the tissues be estimated. These parameters (V, f_b, and f_T) will give greater insight into the interspecies variation in the distribution of radiotracers that distribute beyond the vascular space.

The *blood clearance* for a radiotracer may be defined as the hypothetical volume of blood from which the radiotracer is removed per unit time by all eliminating processes (usually liver plus kidneys). Clearance is expressed in units of volume per time. It is the pharmacokinetic parameter that gives an indication of the *efficiency* of the body's elimination processes for the radiotracer. The blood clearance (CL_b) is calculated from blood disappearance data by dividing the area under the blood radiotracer disappearance curve (C_b versus time) into the injected dose (Wilkinson and Shand, 1975), or,

$$CL_b = \frac{\text{dose}}{AUC} \qquad \text{(Eq. 6-11)}$$

The blood clearance accounts for elimination of radiotracer via all possible routes. Since these routes are usually limited to the liver and/or kidneys, further simple calculations are necessary to obtain renal clearance ($CL_{R,b}$)

and hepatic clearance ($CL_{H,b}$) from the blood clearance. This is possible if the total radioactivity of unchanged or unmetabolized radiotracer excreted in the urine is measured and compared with the injected dose. Then

$$CL_{R,b} = \frac{\text{(radioactivity in urine)}}{\text{dose}} CL_b \qquad \text{(Eq. 6-12)}$$

and

$$CL_{H,b} = CL_b - CL_{R,b} \qquad \text{(Eq. 6-13)}$$

It is important to note that if an accurate estimate of renal clearance is to be made, urine must be collected until excretion ceases. This is usually about six to seven times the longest half-life that can be estimated from the blood disappearance curve. Optimally for hepatobiliary radiotracers renal clearance should be zero and hepatic clearance should approach hepatic blood flow. If the hepatic radiotracer clearance is equal to hepatic blood flow (Q_H), the hepatic extraction ratio (E_H) is equal to 1.0, according to

$$CL_H = Q_H \cdot E \qquad \text{(Eq. 6-14)}$$

Thus when attempting to correlate hepatic radiotracer clearance with physiology in a given animal species, the hepatic blood flow should be known. Values for hepatic (and plasma) blood flow in various species are available (Boxenbaum, 1980), or an estimate may be made from body weight (B) using the heterogonic equation (Adolph, 1949):

$$Q_H = 0.0554B^{0.894} \qquad \text{(Eq. 6-15)}$$

Values for common laboratory species are given in Table 6-3.

Using data from blood collections in rats and dogs, some of the pharmacokinetics described have been determined for [^{99m}Tc]diethyl-IDA (Table 6-4) (Fritzberg et al., 1982a). While it appears that the specificity is similar in both species as indicated by the values of 4.2 and 5.6% for renal excretion in rats and dogs, respectively, other parameters vary considerably. The hepatic clearance is much lower on a per kilogram basis in dogs. Although the hepatic blood flow differs in the same direction, actual comparison of hepatic clearance to hepatic blood flow indicates an extraction ratio of 0.73 in rats and only 0.30 in dogs. While this suggests limitations in correlating data from one species to another, it does indicate that relative differences may be larger in dogs. Thus dogs may be preferable to rats for the comparison of hepatobiliary agents that have similar kinetics.

The distribution volumes, initial and steady state, differ in the same direction for both species, with those in the rat significantly larger on a weight-adjusted basis. The direction of differences is the same as that observed in a comparison of [^{99m}Tc]diethyl-IDA with other agents with lower clearance values, such as [^{99m}Tc]hydralazine and [^{99m}Tc]8-hydroxyquinoline-7-carboxylate (Fritzberg et al., 1982). Since the complexes with higher clearance had larger volumes of distribution, the relationship of clearance and distribution

TABLE 6-4. Comparative Pharmacokinetic Parameters for [^{99m}Tc]Diethyl-IDA in Rats and Dogs

Parameter	Rats	Dogs
Weight (kg)	0.386 ± 0.042	21.5 ± 1.1
Dose in urine, %	4.2[a]	5.6
Hepatic blood flow (ml/min kg)[b]	69.1	41.0
Hepatic clearance (ml/min kg)	50.7 ± 15.7	12.1 ± 2.30
Extraction ratio	0.73	0.30
Blood volume (ml/kg)[b]	64.3	86.1
Initial dilution volume (ml/kg)	515 ± 117	208 ± 53.3
Steady-state distribution volume (ml/kg)	1820 ± 320	751 ± 158

[a] From product information, ^{99m}Tc-SOLCO HIDA, SOLCO Ltd., Basel.
[b] From Table 5-3.

volumes as discussed earlier appears to be inconsistent. The basis seems to be that instead of the distribution volumes simply reflecting dilution into extravascular tissues, the rapid hepatic clearance effectively results in a large distribution volume, since the values obtained are based on measurements of the vascular compartment only. Thus interpretation of the resulting volume parameters requires caution similar to that expressed earlier in regard to rate-limiting steps in excretion of the radiotracer (see p. 189).

LABORATORY ANIMAL TECHNIQUE CONSIDERATIONS

Animal handling techniques can affect results, and several aspects are important to consider in experimental design. Variables such as whether or not the animal was anesthetized, the anesthetic used, temperature maintained during study, and surgical techniques can all affect the values obtained. If studies include, for example, simultaneous administration of [^{131}I]rose bengal with a ^{99m}Tc-labeled radiotracer under study, the results can be interpreted with respect to [^{131}I]rose bengal as the control, and the degree of confidence in the meaningfulness of the results is significantly increased. Direct comparison of data from one laboratory to another may be otherwise difficult.

Anesthetization and Liver Function

Studies in animals should provide results that are relevant to the behavior of radiopharmaceuticals under normal conditions and in the case of disease models the behavior of the radiopharmaceutical as one would expect it to behave in patients with disease. A factor that must be considered in the use of animals is the anesthetic condition that is imposed to facilitate injections, to minimize movement when imaging, and to allow surgical techniques that are necessary for rapid blood sampling, fluid supply, and bile sampling. When the ideal condition of using an unanesthetized animal is impossible, changes in body functions that are induced by the anesthetic must be considered.

A clear effect on hepatobiliary excretion has been observed with body temperature changes. When anesthesia of mice and rats was induced with pentobarbital (Nembutal), body temperature fell when external heat was not supplied (Roberts et al., 1967). The transport maxima (Tm) of bilirubin decreased 5%/°C decrease in rectal temperature in rats and 4%/°C in mice, while the transport maximum (Tm) of BSP decreased 7%/°C in rats. Bile flow decreased about 7%/°C in rats and 8%/°C in mice. Temperature decreases in the animals were 1°C in mice, and 2°C in rats in 1 hr in a room temperature of 23–25°C. Thus it is important to maintain the body temperature of animals under anesthesia using external heat as necessary.

Another consideration is the effect of anesthetics on blood supply to an organ. The relationship between liver blood flow and total hepatic clearance is shown in Fig. 6-2 (Wilkinson and Shand, 1975), as discussed on p. 191. Hepatic clearance is given by the expression

$$Cl_H = Q\frac{[C_A - C_V]}{C_A} = QE \qquad \text{(Eq. 6-16)}$$

where Q was defined as the total hepatic blood flow. The extraction efficiency (E) is expressed by $C_A - C_V/C_A$, where C_A is the mixed hepatic arterial and portal venous concentration of radiotracer supplied to the liver, and C_V is the hepatic venous concentration of radiotracer leaving the liver. If the extraction efficiency is low, there is little dependence of clearance on blood flow. However, if the extraction efficiency is high, the hepatic clearance will be highly dependent on blood flow to the liver. Estimates of extraction efficiencies of several radiopharmaceuticals in humans are 48–56% for [^{99m}Tc]diethyl-IDA (Henrikson and Winkler, 1978), 60% for [^{99m}Tc]N-(2,6,-diisopropylacetanilide)iminodiacetate (Klingensmith et al., 1981), and 66% for [^{99m}Tc]N-(3-bromo-2,4,6-trimethylacetanilide)iminodiacetate (Klingensmith et al., 1982). Values are even higher in rats as seen in the section on pharmacokinetic parameters and as will be discussed under the section on the use of the isolated perfused liver. Thus the researcher must concern himself with the conditions that might alter blood supply to the liver when studying recently developed hepatobiliary radiopharmaceuticals.

Hepatic blood supply has been shown to be significantly altered by anesthesia, but the effect is dependent on the anesthetic used. In a study of hepatic blood flow in patients the use of thiopental-nitrous oxide or cyclopropane reduced the blood supply by about 35% (Levy et al., 1961) while the use of pentobarbital and atropine resulted in no significant change. How-

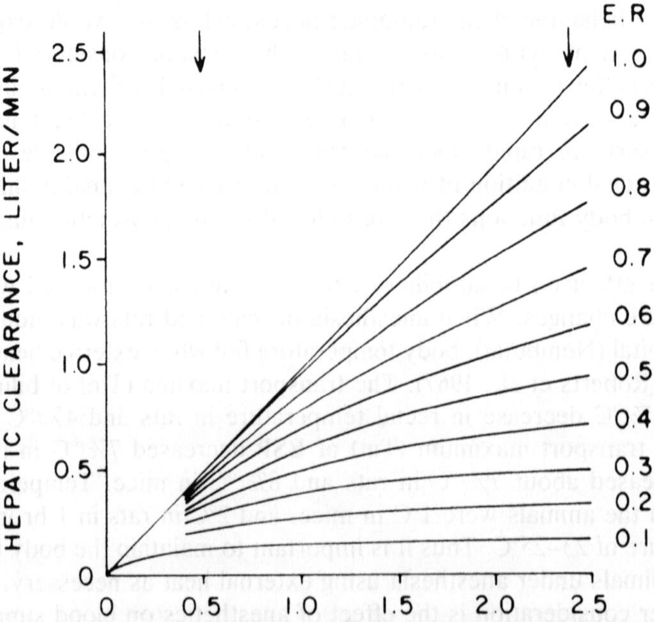

FIG. 6-2. The relationship between liver blood flow and total hepatic clearance for drugs with varying extraction ratios (ER) or fraction extracted by the liver. The arrows indicate the normal physiological range of liver blood flow and the extraction values refer to a normal flow of 1.5 liter/min (Wilkinson and Shand, 1975).

ever, the doses given patients were much lower than those commonly used in animal studies, and correlations to animal dose levels would not be reasonable.

These results are consistent in any case with a study on the effects of various anesthetic agents on hepatobiliary excretion parameters in rats (Cooper et al., 1976). In the experiments described, the common bile duct of rats was cannulated, and then various parameters such as bile flow, bile salt and bilirubin excretion, and enzyme activity in the bile were determined for five days postsurgery. After stabilization several anesthetic conditions were imposed and effects studied. Ether and halothane caused a decrease in bile flow rate and bile salt and bilirubin excretion. Pentobarbital injection at 40 mg/kg did not produce a significant change in the preceding parameters, although a tendency for a decrease in values for the parameters was observed. Similarly, β-glucuronidase and acid phosphatase in bile were increased with ether and halothane, but not with pentobarbital. The results reported indicate concern for the type of anesthetic used as well as the dose. For the rat at least, pentobarbital (Nembutal) appears to be the agent of choice.

A study of hepatic uptake of cortisol in conscious and anesthetized sheep indicated that the use of pentobarbital caused changes in hepatic blood flow

(Paterson and Harrisonn, 1972). Hepatic plasma flow based on BSP measurements in the anesthetized sheep was only one third of that in the conscious sheep. Thus species generalizations of the effect of anesthetic agent are not safe and control studies are required.

In rabbits the range between sedation and lethal doses of pentobarbital is small. Procedures that require sedation are done in our laboratory by anesthetization with 9/mg/kg of xylazine followed in 10 min by 45 mg/kg ketamine.

The fact that the use of pentobarbital in the rat or the optimum anesthetic in other species may result in at least small differences emphasizes the importance of using a control radiopharmaceutical such as [^{131}I]rose bengal for hepatobiliary studies. Differences in [^{131}I]rose bengal behavior indicate undesired changes in the physiological state of the animal. When it is not practical to use such a control radiopharmaceutical, the researcher must control the dose level of the anesthetic, include other controls, and do a sufficient number of animals so that the results are statistically significant.

When animals such as rabbits or dogs may be used repeatedly for experiments that require anesthesia, increased bile flow and transport capacity must be considered. Phenobarbital has been shown to induce changes in the liver that affect these parameters (Roberts and Plaa, 1967). For example, the transport maximum of bilirubin was increased 32%, and bile flow increased 61% by treatment of rats with 60 mg/kg IP phenobarbital once daily for 3 days. While comparable results with pentobarbital are not available, similar changes should be considered when multiple frequent studies are carried out on the same animals.

Species differences in gallbladder activity exist. For example, anesthetized rabbits show transient gallbladder activity in imaging studies, but the gallbladder reflex of dogs is slowed to the point that only gallbladder is seen at times well beyond 1 hr postinjection (Fig. 6-3). In the examples shown in Fig. 6-4 there are two differences in the rabbit images that are apparent. Gut motility seems to be slowed in the anesthetized (xylazine/ketamine) rabbit such that gut activity is much slower in reaching the small intestine. There seems to be faster blood clearance in the anesthetized rabbit. These differences would need to be evaluated quantitatively in larger numbers to verify significance but are likely examples of differences resulting from the anesthetic treatment. Although intermediate images were not available, the unanesthetized dog showed the major portion of the radioactivity in the gut at 60 min, while the anesthetized dog showed no lower gut radioactivity at 2 hr. The absence of lower gut activity at times less than 1 hr in anesthetized dogs has been typical.

Age Dependency of Hepatic Function

Hepatic function in rats has been shown to change with age (Fischer et al., 1979). The transport of eosine, an anion, was highest in 60-day-old rats and

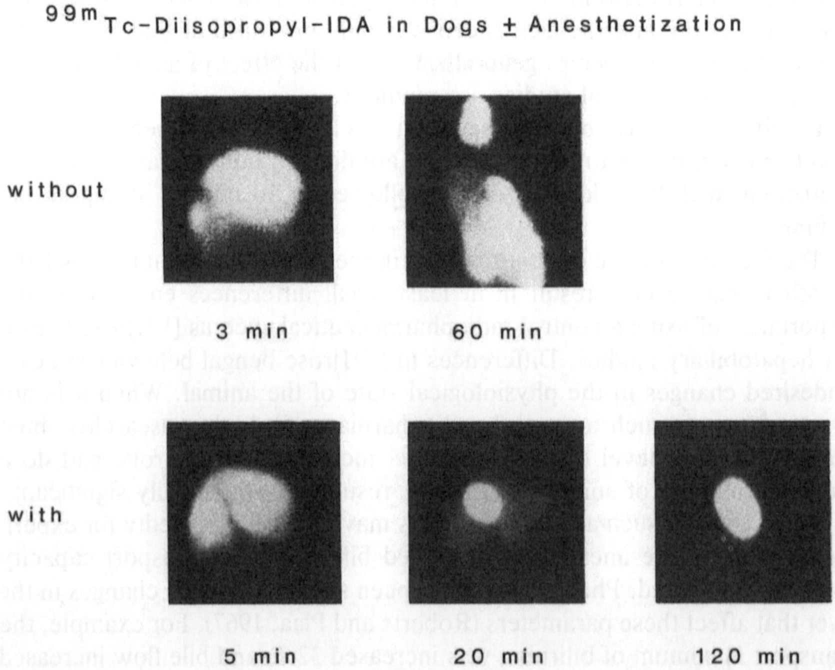

FIG. 6-3. Comparison of [^{99m}Tc]diisopropyl-IDA in unanesthetized and anesthetized (pentobarbital) dogs. Clearance is rapid in both conditions, but gallbladder reflex is very slow. Typically gut activity beyond the gallbladder was seen only after 60 min in dogs under anesthetization.

then decreased to about 70% by 120 days. Bile flow was highest at 30 days and then decreased by 25% by 60 days and 35% by 120 days. These days correspond generally to about 100 g for 30-day-old rats, 250 g for 60-day-old rats, and 375 g for 80-day-old rats.

Surgical Techniques for Blood and Bile Sampling

Because of size, cost, and a long history of pharmacological studies, the rat is the classic animal model when kinetic studies requiring blood and bile samples are carried out. The methodology that we follow was described in detail by Fritzberg et al. (1979). Rats weighing 250–350 are anesthetized with 50 mg/kg doses of sodium pentobarbital IP. Later 3–5-mg doses are given to maintain the desired level of anesthesia. Injections and blood and bile sampling are made through polyethylene tubing (PE 10). This tubing is inserted into PE 50 tubing in order to minimize backflow pressure. The tubing for injections and hydration is placed in a femoral vein and the tubing for blood sampling is positioned in the descending aorta via the left carotid artery. Blood sample volumes should be minimized so that total blood volume re-

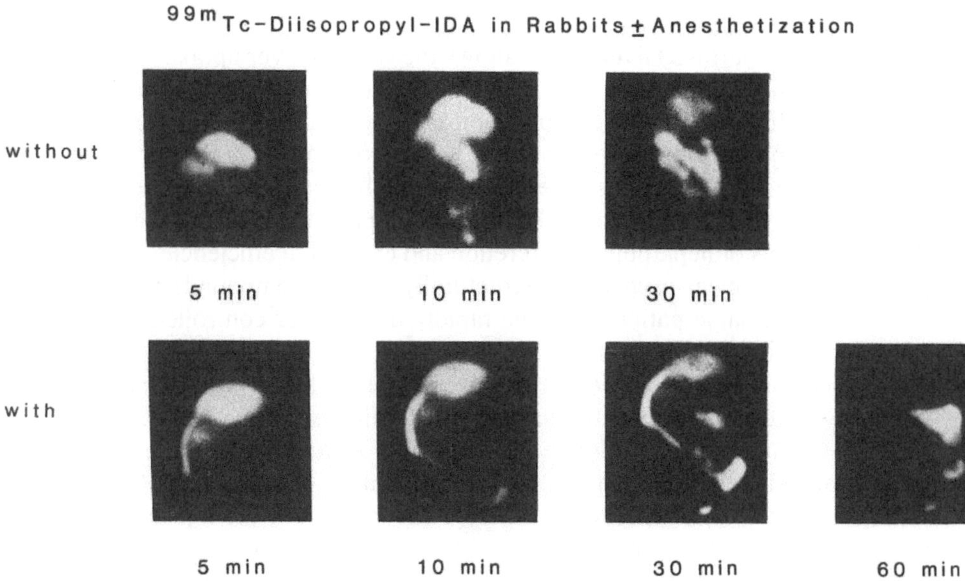

FIG. 6-4. Comparison of time course of [^{99m}Tc]diisopropyl-IDA in unanesthetized and anesthetized (xylazine/ketamine) rabbits. The clearance from the blood is similar in both conditions as seen by the absence of cardiac blood pool activity at 5 min postinjection. Otherwise the striking difference is the slow gut motility seen in the anesthetized rabbit.

moved from the animal does not cause shock. Bile is collected by placing the cannula in the common bile duct. The tubing can be connected to a small fraction collector for easy timed collection of bile samples, or when samples of short time duration are desired, the samples can be manually collected on absorbent paper or in tubes containing some water to minimize carryover from one sample to the next. If chronic studies are planned the bile fistula can be placed such that it leaves the body through skin on the back of the animal (Cooper et al., 1976). Bile flow under these conditions is about 25 μl/min 300 g rat.

The Perfused Liver Model

Perfused livers have been used for studies of liver physiology for a number of years (Miller, 1973; Reichen and Paumgartner, 1976). The procedure involves perfusing the liver of the animal with an oxygenated medium that contains buffer, glucose, and optimally erythrocytes (Hanks et al., 1980). The pH of the perfusate, the temperature of the cabinet, and the pressure of the perfusate are all controlled. The liver is left attached to the animal and the perfusate is connected to the portal vein and hepatic artery. The outflow of the liver through the hepatic vein is recirculated or can be collected for measurements. The common bile duct can be cannulated for bile output measurements.

The isolated perfused liver model allows the study of liver uptake, metabolism, and biliary excretion without the complications of clearance by other organs and pharmacokinetic complications. Competition for hepatic uptake is controlled by the presence or absence of such compounds in the perfusate medium. The liver functions essentially normally with structural components such as membranes intact. Thus it is a simplified model for studies of the mechanisms of hepatobiliary excretion and extraction efficiencies. Comparative kinetic measurements of structurally related compounds that are handled by the same pathway can be rapidly done under controlled conditions and via direct measurements.

Fritzberg and Reichen (1983) recently measured the hepatic extraction efficiency of a series of hepatobiliary agents by the use of the isolated perfused liver model. The radiopharmaceutical was added to infusate just before it reached the portal vein, and the hepatic venous outflow was collected for 3–5 min afterward in order to measure the fraction of the injectate not extracted by the liver. Fractionation of the collected blood indicated that less than 1% was in the 3–5-min portion. The extraction efficiency was calculated by the expression

$$\text{Extraction efficiency (\%)} = 100 - \left(\frac{\text{counts (blood)}}{\text{counts injected}} \times 100\right) \quad \text{(Eq. 6-17)}$$

Cannulation of the common bile duct was also done to measure bile output kinetics and hepatocyte transit times. The results are shown in Table 6-5 and Fig. 6-5.

TABLE 6-5. Hepatobiliary Radiopharmaceutical Parameters from the Isolated Perfused Rat Liver

Radiopharmaceutical	*Extraction Efficiency (%)*	T_{50} *(min)*[a]	*Slope (%/min)*[b]
[^{99m}Tc]diethyl-IDA	72.8 (69.3–78.6)	3.5 (2.8–3.8)	−1.85
[^{99m}Tc]diisopropyl-IDA	79.1 (74.0–86.6)	2.9 (2.8–3.0)	−2.25
[^{99m}Tc]*p*-isopropyl-IDA	58.9 (54.7–62.2)	4.7 (3.8–5.5)	−0.74
[^{99m}Tc]*p*-butyl-IDA	90.5 (89.4–91.6)	9.0 (8.0–10.0)	−0.22
[^{99m}Tc]trimethylbromo-IDA	74.1 (71.9–78.2)	6.8 (6.0–7.6)	−0.42
[^{99m}Tc]PHMT[c]	92.1 (89.9–95.4)	3.0 (2.8–3.3)	−2.23
[^{131}I]rose bengal	12.3 (5.3–20.6)	>30	—

[a] Time for peak bile percentage collected in 1 min to decrease by one-half.
[b] Downslope of bile output curve from 1-min samples.
[c] *N*-Pyridoxylmethyltryptophan.

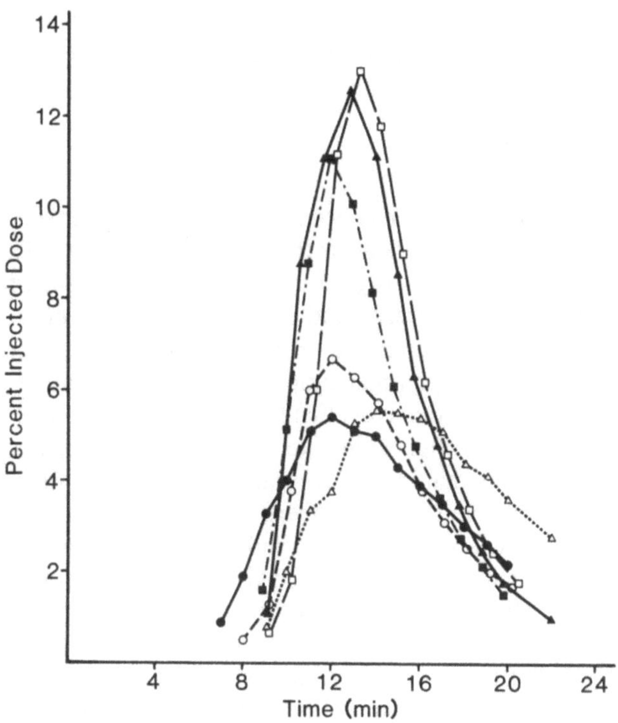

FIG. 6-5. Comparative rates of appearance of hepatobiliary agents in the bile in the isolated perfused rat liver system. Each data point represents the percent of the administered dose collected in 1 min.

The trend of the results is in general as expected. In these studies the extraction efficiency of [^{99m}Tc]diisopropyl-IDA is slightly higher than [^{99m}Tc]trimethylbromo-IDA, while in humans the reverse was true. In both cases the differences are small and are not statistically significant. The new Schiff-base-derived agent, [^{99m}Tc]*N*-pyridoxyl-5-methyltryptophan ([^{99m}Tc]PHMT) (Kato-Azuma, 1982) resulted in an average extraction efficiency of over 90%, and thus was comparable to taurocholate (Iga and Klaassen, 1982; Pries et al., 1981). Surprisingly, [^{99m}Tc]*p*-butyl-IDA also resulted in extraction efficiency values of about 90%.

Biliary excretion curves based on percent dose collected in 1-min samples (Fig. 6-5) were also in general as expected. The appearance and disappearance of [^{99m}Tc]*p*-isopropyl-IDA and [^{99m}Tc]*p*-butyl-IDA were slower than [^{99m}Tc]diethyl-IDA and [^{99m}Tc]diisopropyl-IDA. These results correlate with clinically observed differences in hepatocyte transit times. The rapid excretion of [^{99m}Tc]PHMT was as expected from rabbit and whole rat data (Kato-Azuma, 1982). The curve for [^{99m}Tc]trimethylbromo-IDA, however, indicates slower biliary excretion than that expected from rabbit results

(Nunn et al., 1981) and clinical results (Klingensmith et al., 1982). The results of [^{99m}Tc]trimethylbromo-IDA suggest species differences and caution in extrapolation from rat data only.

Isolated Hepatocyte Uptake Studies

Another simplified model that has been used for the comparison of different hepatobiliary radiopharmaceuticals is the isolated hepatocyte. The cell preparation is carried out by perfusing rat livers with collagenase, which digests the cell-adhering material (Berry and Friend, 1969; Schwenk, 1980). The hepatocytes can then be separated from each other and resuspended in buffer containing the radiopharmaceutical and other compounds of interest.

This method has been used to evaluate relative rates of uptake of potential hepatobiliary agents and their ability to be taken up by hepatocytes in the presence of compounds that may share transport pathways such as bilirubin and bile acids (Chervu et al., 1979; Loberg et al., 1981). The hepatocytes are incubated with the hepatobiliary agent under study and cell binding or uptake measured by centrifugation of the cells through a silicone layer into a trapping medium. The density of the silicone oil is less than that of the cells, and the incubation medium density is less than that of the silicone oil. Results with [^{99m}Tc]IDA complexes have shown differences in uptake between these agents over 30 min of incubation (Loberg et al., 1979). The addition of compounds such as sulfobromophthalein (BSP), bilirubin, and taurocholate also resulted in the inhibition of uptake of these agents (Loberg et al., 1979). The result with taurocholate is in contrast to studies in rats in which the excretion of [^{99m}Tc]diethyl-IDA and [^{131}I]rose bengal was increased by taurocholate, presumably due to a choleretic effect (Fritzberg et al., 1979). The cationic quarternary amine, oxyphenonium did not inhibit hepatocyte uptake in agreement with studies in rats (Fritzberg et al., 1979) and dogs (Harvey et al., 1979). These results indicate the utility of this method for comparing agents as well as for the study of mechanisms of hepatocyte transport of these agents.

In general the use of hepatocytes has given results in good agreement with those obtained in the isolated perfused liver. Thus transport systems seem to be preserved (Loberg et al., 1979). Advantages of the isolated hepatocytes are the absence of hemodynamic differences, the potential repetition of experiments using the cells of one animal, the ability to subject the cells to extreme conditions of pH, oxygen pressure, and the absence of other cell types, such as Kupfer cells.

Isolated hepatocytes are different, however, from cells in the intact liver, since the hepatocyte has two different surfaces. One is in contact with the sinusoidal compartment and hence blood plasma, and the other is in contact with bile. One is oriented toward uptake that is often reversible, and the other is oriented toward excretion. Both membrane surfaces are exposed, however, in the suspended isolated hepatocyte. These considerations can complicate comparative uptake interpretations and their translation to functioning liver results.

Manipulations of Animals to Model Hepatobiliary Disease

Several attempts have been made to simulate clinically relevant disease conditions in animal models. For hepatobiliary agents two major areas are involved in clinical interpretations. One is the effect of bile duct blockage and bile flow problems on these agents; the other is the effect of decreased levels of hepatocyte function. Often hepatocellular function is decreased, and the question is whether the decrease is primary or secondary to bile flow problems. Imaging of the biliary tree and demonstration of biliary excretion is then necessary for diagnosis.

The question "What is the effect of bile tract obstruction on the behavior of hepatobiliary agents?" can be expanded to "What is the time course of biliary tract obstruction effect on these agents?" This was studied in dogs by surgically ligating the common bile duct and studying the hepatic clearance and bile flow as indicated by gallbladder filling with [^{99m}Tc]diethyl-IDA (Klingensmith et al., 1981). The common bile duct was ligated below the gallbladder. Thus as long as the gallbladder was able to reabsorb water, bile flow would be indicated by uptake of radiopharmaceutical by the gallbladder. Although results clearly indicated the lack of bile flow and decreased hepatic clearance by the third day postoperation (Fig. 6-6), some hepatocyte function persisted for weeks. Thus the time required for complete obstruction to produce severe hepatocyte damage in patients could not be estimated from the study.

Bilirubin, the metabolic product of heme degradation, is eliminated from the body by biliary excretion. As hepatocyte function decreases, levels of plasma bilirubin increase. Since studies have shown that other members of the bilirubin pathway such as BSP in dogs (Harvey et al., 1979) and rats (Fritzberg et al., 1979), indocyanine green in rats (Fritzberg, 1982), and bilirubin itself in rabbits (Jansholt et al., 1979) decrease the rate of hepatic uptake and biliary excretion of [^{99m}Tc]iminodiacetates, these agents share the same pathway and their uptake and excretion may be expected also to be retarded in patients with increased levels of bilirubin. Thus animal studies using high levels of compounds that share the pathway of bilirubin can be expected to mimic the jaundiced clinical condition. However, the jaundiced patient or animal not only has increased bilirubin levels but usually has underlying decreased levels of hepatocellular function. However, studies on animals with liver pathology may be more useful in predicting behavior in the clinical situation.

Although many agents have been used to cause liver toxicity as models for human liver disease, few reports have appeared that describe such effects on hepatobiliary agents. A variety of agents, including cirrhosis induced by carbon tetrachloride and phenobarbital (McLean et al., 1969) and

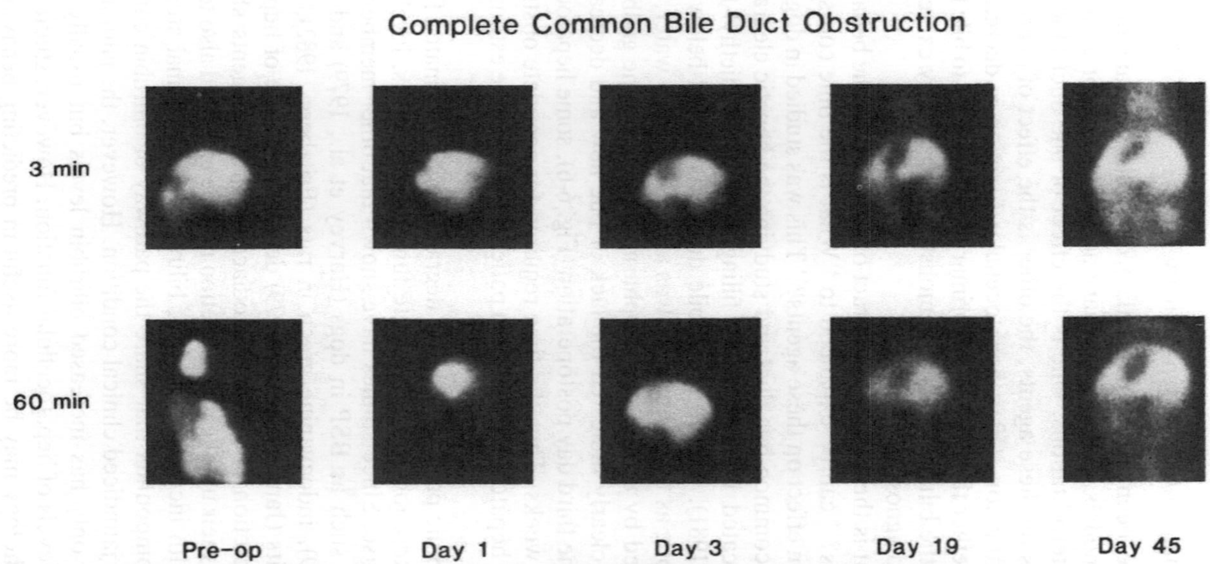

FIG. 6-6. Ventral images at 3 and 60 min after injection of [^{99m}Tc]diethyl-IDA shown for preoperative control and postoperative studies on days 1, 3, 19, and 45. Control study shows good hepatocyte clearance, namely, high ratios for liver to background and liver to cardiac blood pool at 3 min and passage of bolus into intestine and gallbladder by 60 min. Study at 3 hr after operation shows good hepatocyte clearance, visualization of dilated gallbladder, and no intestinal radioactivity. All subsequent studies show decreased hepatocyte clearance, nonvisualization of biliary tract (except as areas of decreased radioactivity), and no intestinal radioactivity (McLean et al., 1969).

cholestasis by α-napthylisothiocyanate, estrogen derivatives, or phalloidin have been reviewed (Zimmerman, 1978).

We have performed initial extraction efficiency determinations on cirrhotic rats from carbon tetrachloride and phenobarbital treatment using the perfused liver model (Fritzberg and Reichen, unpublished). The extraction efficiency of the treated rats of 25 and 14% for [^{99m}Tc]diisopropyl-IDA, is markedly reduced from the normal value of 79%.

SUMMARY

The development of hepatobiliary agents has come a long way from the time when some biliary radioactivity was the basis for the claim of a "gallbladder imaging agent" to recent times when agents with high specificity for biliary excretion, high hepatic extraction efficiency, and short hepatocyte transit times have been characterized. Correspondingly, sophistication in the use of animals for evaluation of these agents has also progressed. The progression may be described as going from a time course of organ radioactivity content in mice and a time course of gamma camera images in rabbits or dogs to methods for collection of bile in rats (Fritzberg et al., 1979), dogs (Harvey et al., 1979), and baboons (Wistow et al., 1977) that made possible quantitative comparisons of biliary excretion rates. The techniques of blood and bile collections in rats and dogs were extended (Fritzberg et al., 1979; Harvey et al., 1979) to studies of the *in vivo* kinetics of the hepatobiliary agents evaluated in the presence of saturating levels of compounds known to be transported by the different pathways of hepatocytes in order to answer questions of hepatocyte transport mechanisms and ability to compete for hepatic uptake and biliary excretion in patients with elevated levels of serum bilirubin. Simplified models of these types of studies were provided by the isolated perfused liver (Miller, 1973; Reichen and Paumgartner, 1976; Fritzberg and Reichen, 1983) and isolated hepatocytes (Loberg and Porter, 1979). By appropriate choice of these models one can define the behavior of a hepatobiliary radiopharmaceutical as well as provide a quantitative comparison of it to the standard(s) of the time.

The development of models for disease conditions is at a comparatively early stage. The results of obstruction (Klingensmith et al., 1981) and cirrhosis models (Fritzberg and Reichen, unpublished) described in this review represent early stage studies of this type. A significant amount of development of models of cirrhosis and cholestasis has been described (Zimmerman, 1978), and application to radiopharmaceutical behavior in these models will likely be reported in the near future.

Is there a perfect animal model for hepatobiliary agent development? The answer is not obvious from reviewing the studies in the literature. The extrapolation of pharmacokinetics from one animal to another or to man would not appear to be straightforward as seen by our comparison of [^{99m}Tc] diethyl-IDA parameters in rats and dogs. However, in general, the trend of differences carries over from one species to another. Thus by doing studies that are well defined quantitatively and by making comparisons with well-defined reference compounds that are similarly transported by the hepatocytes the relative behavior in other species or in humans can be inferred with some confidence. The slow hepatocyte transit of [^{99m}Tc]trimethylbromo-IDA in rats and rapid transit in rabbits and humans indicate caution in extrapolations to man from studies in only one species, however. The species chosen may not be as important as the ability to determine detailed quantitative kinetic data that define the rates of transport, as well as the specificity for hepatobiliary excretion.

REFERENCES

Adolph, E. F. (1949). Quantitative relations in the physiological constitutions of mammals. *Science* **109**:579–585.

Alpert, S., Mosher, M., and Shanske, A. (1969). Multiplicity of hepatic excretory mechanisms for organic anions. *J. Gen. Physiol.* **53**:288–247.

Altman, P. L., and Dittmer, D. S. (1972a). *Biology Data Book*. Bethesda: Federation of American Societies for Experimental Biology, pp. 1703–1706.

Altman, P. L., and Dittmer, D. S. (1972b). *Biology Data Book*. Bethesda: Federation of American Societies for Experimental Biology, pp. 1841–1849.

Arias, I. M. (1972). Transfer of bilirubin from blood to bile. *Sem. Hematol.* **9**:55–70.

Arias, I. M., Fleishner, G., Kirsch, R., Mishkin, S., and Gatmaitan, F. (1976). *Glutathione: Metabolism and Function,* Arias, I. M., Jakoby, W. B., eds. New York: Raven Press, pp. 175.

Arias, I. M., Gartner, L. M., Cohen, M., Ben-Ezzer, J., and Levi, A. J. (1969). Chronic nonhemolytic unconjugated hyperbilirubinemia with glucuronyl transferase deficiency. *Am. J. Med.* **47**:395–409.

Aziz, F. T. A., Hirom, P. C., Millburn, P., Smith, R. L., and Williams, R. T. (1971). The biliary excretion of anions of molecular weight 300–800 in the rat, guinea pig and rabbit. *Biochem. J.* **125**:25–26P.

Baker, R. J., Bellen, J. C., and Ronai, P. M. (1975). Technetium ^{99m}Tc-pyridoxylideneglutamate: A new hepatobiliary radiopharmaceutical. 1. Experimental aspects. *J. Nucl. Med.* **16**:720–727.

Benet, L. Z., and Galeazzi, R. L. (1979). Noncompartmental determination of the steady-state volume of distribution, *J. Pharm. Sci.* **68**:1071–1074.

Benet, L. Z., and Ronfeld, R. A. (1969). Volume terms in pharmacokinetics, *J. Pharm. Sci.* **58**:639–641.

Berry, M. N., and Friend, D. S.: (1969). High yield preparation of isolated rat liver parenchymal cells. *J. Cell. Biol.* **43**:405–520.

Billing, B. H., Williams, R., and Richards, T. G. (1964). Defects in hepatic transport of bilirubin in congenital hyperbilirubinemia: An analysis of plasma bilirubin disappearance curves. *Clin. Sci.* **27**:245–257.

Black, M., and Billing, B. H. (1969). Hepatic bilirubin UDP-glucuronyl transferase activity in liver disease and Gilbert's syndrome. *N. Engl. J. Med.* **280:**1266–1271.

Boxenbaum, H. (1980). Interspecies variation in liver weight, hepatic blood flow, and antipyrine intrinsic clearance: Extrapolation of data to enzodiazepines and phenytoin, *J. Pharmacol. Biopharmacol.* **8:**165–176.

Chervu, L. R., Nunn, A. D., and Loberg, M. D. (1982). Radiopharmaceuticals for hepatobiliary imaging. *Sem. Nucl. Med.* **12:**5–17.

Chervu, L. R., Robbins, E. B., Huq, S. S., and Blaufox, M. D. (1979). In vivo and in vitro studies of Tc-99m HIDA uptake. *J. Nucl. Med.* **20:**655.

Chiotellis, E., and Varvarigou, A. (1980). ^{99m}Tc-labeled *N*-substituted carbamoyl iminodiacetates: Relationship between structure and biodistribution. *Int. J. Nucl. Med. Biol.* **7:**1–7.

Cooper, B., Eakins, M. N., and Slater, T. F. (1976). The effect of various anesthetic techniques on the flow rate, constituents and enzyme composition of rat bile. *Biochem. Pharmacol.* **25:**1711–1718.

Dhumeaux, D., Berthelot, P., and Javitt, N. B. (1974). Dibromosulfophthalein (DBSP) estimation of hepatic transport function in man. *Eur. J. Clin. Invest.* **4:**181–185.

Dubin, I. N., and Johnson, F. B. (1954). Chronic idiopathic jaundice with unidentified pigment in liver cells. *Medicine* **33:**155–197.

Fevery, J., Leroy, P., Van De Vijver, M., and Heirwegh, K. P. M. (1972). Enzymic transfer of glucose and xylose from uridine diphosphate glucose and uridine diphosphate xylose to bilirubin by untreated and digitonin-activated preparations from rat liver. *Biochem. J.* **129:**619–633.

Fischer, E., Barth, A., Varga, F., and Klinger, W. (1979). Age dependence of hepatic transport in control and phenobarbital-pretreated rats. *Life Sci.* **24:**557–562.

Fritzberg, A. R. (1981). The evaluation of hepatocyte function with radiotracers. In *Studies of Cellular Function Using Radiotracers.* M. W. Billinghurst, ed., Boca Raton, Fla.: CRC Press, pp. 73–92.

Fritzberg, A. R., Bloedow, D. C., Klingensmith, W. C., and Whitney, W. P. (1982a). Comparative study of ^{99m}Tc-hepatobiliary agents based on naphthalene and similar ring systems. *Int. J. Nucl. Med. Biol.* **9:**1–11.

Fritzberg, A. R., and Klingensmith, W. C. (1982). Teaching editorial: Quest for the perfect hepatobiliary radiopharmaceutical. *J. Nucl. Med.* **23:**543–546.

Fritzberg, A. R., Klingensmith III, W. C., Whitney, W. P., and Kuni, C. C. (1981). Chemical and biological studies of Tc-99m *N,N'*-bis(mercaptoacetamido)ethylenediamine: A potential replacement for I-131 iodohippurate. *J. Nucl. Med.* **22:**258–263.

Fritzberg, A. R., Kuni, C.C., Klingensmith, W. C., Stevens, J., and Whitney, W. P. (1982b). Synthesis and biological evaluation of Tc-99m *N,N'*-bis(mercaptoacetyl)-2,3-diaminopropanoate: A potential replacement for [^{131}I]*o*-iodohippurate. *J. Nucl. Med.* **23:**592–598.

Fritzberg, A. R., and Reichen, J. (1983). Direct measurement of extraction efficiencies and mean transit times of hepatobiliary agents in the perfused rat liver. *J. Nucl. Med.,* **24:**126.

Fritzberg, A. R., Whitney, W. P., and Klingensmith III, W. C. (1979). Hepatobiliary transport mechanism of Tc-99m diethyl-IDA. In *Radiopharmaceuticals II,* Sodd, V. J., Allen, D. R., Hoogland, D. R., Ice, R. I., eds. New York: Society of Nuclear Medicine, pp. 566–586.

Frizzell, R. A., and Heintze, K. (1980). Transport functions of the gallbladder. In *Liver and Biliary Tract Physiology I,* Int. Rev. Physiol., Vol. 21, Javitt, N. B., ed. Baltimore: University Park Press, pp. 221–247.

Gelius, L., Skretting, A., and Aas, M. (1981). A mathematical model for the liver uptake and excretion of ^{99m}Tc-diethyl-IDA. *Eur. J. Nucl. Med.* **6:**139–142.

Gibaldi, M., and McNamara, P. J. (1978). Apparent volumes of distribution and drug binding to plasma proteins and tissues. *Eur. J. Clin. Pharmacol.* **13:**373–378.

Gibaldi, M., and Perrier, D. (1975a). *Pharmacokinetics*. New York: Marcel Dekker, pp. 175–187.

Gibaldi, M., and Perrier, D. (1975b). *Pharmacokinetics*. New York: Marcel Dekker, p. 315.

Gillette, J. R. (1971). Factors affecting drug metabolism. *Ann. N.Y. Acad. Sci.* **179:**43–66.

Goresky, C. A. (1980). Uptake in the liver. The nature of the process. In *Liver and Biliary Tract Physiology I,* Int. Rev. Physiol., Vol. 21, Javitt, N. B., ed. Baltimore: University Park Press, pp. 65–101.

Goresky, C. A., and Kleiger, S. W. (1969). The relation between bile flow and transport maximum for bilirubin in the dog. *Gasteroenterology* **56:**398.

Habig, W., Pabst, M., Fleishner, G., Gatmaitan, F., Arias, I. M., and Jakoby, W. (1974). The identity of glutathione transferase B with ligandin, a major binding protein of the liver. *Proc. Natl. Acad. Sci. USA* **71:**3879–3882.

Hanks, J. B., Meyers, W. C., Willman, C. L., Hill, R. C., and Jones, R. S. (1980). The effect of cell-free and erythrocyte-containing perfusion in rat livers. *J. Surg. Res.* **24:**149–160.

Harvey, E., Loberg, M., Ryan, J., Sikorski, S., Faith, W., and Cooper, M. (1979). Hepatic clearance mechanism of Tc-99m-HIDA and its effect on quantitation of hepatobiliary function. *J. Nucl. Med.* **20:**310–313.

Henrikson, J. H., and Winkler, K. (1978). Pharmacokinetics of ^{99m}Tc-diethyl-IDA in man. In *Proceedings of Symposium on Hepatobiliary Scintigraphy by Means of IDA Derivatives*. Biersack, H. J., Maxsted, J., eds. Darmstadt: G.I.T. Verlag E. Giebeler, pp. 55–70.

Herz, R., Cueni, B., Bircher, J., and Paumgartner, G. (1973). The excretory capacity of the isolated perfused rat liver. *Naunyn Schmiedebergs Arch. Pharmacol.* **277:**297–304.

Hirom, P. C., Millburn, P., and Smith, R. L. (1976). Bile and urine as complementary pathways for the excretion of foreign compounds. *Xenobiotica* **6:**55–64.

Iga, T., and Klaassen, C. D. (1982). Hepatic extraction of bile acids in rats. *Biochem. Pharmacol.* **31:**205–209.

Jansholt, A.-L., Vera, D. R., Krohn, K. A., and Stadalnik, R. C. (1979). In vivo kinetics of hepatobiliary agents in jaundiced animals. In *Radiopharmaceuticals II,* Sodd, V. J., Allen, D. R., Hoogland, D. R., Ice, R. D., eds., New York: Society of Nuclear Medicine, pp. 555–564.

Jones, A. B., Davison, A., LaTegola, M. R., Brodack, J. W., Orvig, C., Sohn, M., Toothaker, A. K., Lock, C. J. L., Franklin, K. J., Costello, C. E., Carr, S. A., Biemann, K., and Kaplan, M. L. (1982). Chemical and in vivo studies of the anion oxo[*N*,*N'*-ethylenebis(2-mercaptoacetimido)]technetate(V). *J. Nucl. Med.* **23:**801–809.

Kato-Azuma, M. (1981). Identification of a mixed ligand complex of technetium-

99m: A chromatographic approach to the chemical structure of carrier free technetium-99m (Sn) pyridoxylideneaminate. *Int. J. Appl. Rad. Isot.* **32:**187–189.

Kato-Azuma, M. (1982). Tc-99m(Sn)-*N*-pyridoxylaminates: A new series of hepatobiliary imaging agents. *J. Nucl. Med.* **23:**517–524.

Kato, M., and Hazue, M. (1978). Tc-99m-(Sn)pyridoxylidene aminates: Preparation and biological evaluation. *J. Nucl. Med.* **19:**397–406.

Klaassen, C. D. (1975). Biliary excretion of xenobiotics. *CRC Crit. Rev. Toxicol.* **4:**1–30.

Klaassen, C. D., and Plaa, G. L. (1967). Species variation in metabolism, storage, and excretion of sulfobromophthalein. *Am. J. Physiol.* **213:**1322–1326.

Klingensmith, W. C., Fritzberg, A. R., Spitzer, V. M., Kuni, C. C., and Shanahan, W. S. M. (1981). Clinical comparison of diisopropyl-IDA Tc-99m and diethyl-IDA Tc-99m for evaluation of the hepatobiliary system. *Radiology* **140:**791–795.

Klingensmith, W., Fritzberg, A., Spitzer, V., Kuni, C., Williamson, M., Gerhold, J., Nunn, A., Loberg, M., and Beisicki, T. (1982). Clinical evaluation of Tc-99m-mebrofenin and comparison with Tc-disofenin for radionuclide hepatobiliary imaging. *Proceedings of World Fed. Nucl. Med. Biol. Mtng.,* Paris, pp. 1596–1598.

Klingensmith, W. C., Gerhold, J. P., Fritzberg, A. R., Spitzer, V. M., Kuni, C. C., Singer, C. J., and Weil, R. (1982). Clinical comparison of Tc-99m-*N,N′*-bis(mercaptoacetamido)ethylenediamine and I-131-hippuran for evaluation of renal tubular function. *J. Nucl. Med.* **23:**377–380.

Klingensmith, W. C., Whitney, W. P., Spitzer, V. M., Klintmalm, G. B. G., Koep, L. M., and Kuni, C. C. (1981). Effect of complete biliary-tract obstruction on serial hepatobiliary imaging in an experimental model. *J. Nucl. Med.* **22:**866–868.

Kurz, H., Trunk, H., and Weitz, B. (1977). Evaluation of methods to determine protein-binding of drugs. *Arzneim-Forsch* **27:**1373–1380.

Levi, A. T., Gatmaitan, Z., and Arias, I. M. (1969). Two hepatic cytoplasmic protein fractions, Y and Z, and their possible role in the hepatic uptake of bilirubin, sulfobromophthalein, and other anions. *J. Clin. Invest.* **48:**2156–2160.

Levy, M. L., Palazzi, H. M., Nardi, G. L., and Bunker, J. P. (1961). Hepatic blood flow variations during surgical anesthesia in man measured by radioactive colloid. *Surg. Gyn. Obstet.* **112:**289–294.

Litwack, G., Ketterer, B., and Arias, I. M. (1971). Ligandin, An abundant liver protein which binds steroids, bilirubin, carcinogens, and a number of exogenous anions. *Nature (Lond.)* **234:**466–467.

Loberg, M. D., Cooper, M., Harvey, E., Callery, P., and Faith, W. (1976). Development of new radiopharmaceuticals based on *N*-substitution of iminodiacetic acid. *J. Nucl. Med.* **17:**633–638.

Loberg, M. D., Nunn, A. D., and Porter, D. W. (1981). Development of hepatobiliary imaging agents. In *Nuclear Medicine Annual 1981,* Freeman, L. M., and Weissmann, H. S., eds., New York: Raven Press, pp. 1–33.

Loberg, M. D., and Porter, D. W. (1979). Review and current status of hepatobiliary agents. In *Radiopharmaceuticals II.*, Eds., Sodd, V. J., Allen, D. R., Hoogland, D. R., and Ice, R. D. New York: Society of Nuclear Medicine, pp. 555–564.

Mia, A. S., Gronwald, R. R., and Cornelius, C. E. (1920). Bilirubin-^{14}C turnover studies in normal and mutant Southdown sheep with congenital hyperbilirubinemia. *Proc. Soc. Exp. Biol. Med.* **133:**955–959.

Miller, L. L. (1973). Technique of isolated rat liver perfusion. In *Isolated Liver*

Perfusion and Its Applications. Bartosek, I., Guaitani, A., Miller, L. L., eds. New York: Raven Press, pp. 11–52.

McLean, E. K., McLean, A. E. M., and Sutton, P. M. (1969). Instant cirrhosis: An improved method for producing cirrhosis of the liver in rats by simultaneous administration of carbon tetrachloride and phenobarbitone. *Br. J. Exp. Pathol.* **50:**502–506.

Nunn, A. D., Loberg, M. D., Conley, R. A., and Schram, E. (1981). The development of a new cholescintigraphic agent, Tc-SQ 26,962, using a structure-distribution relationship approach. *J. Nucl. Med.* **22:**P51.

O'Maille, E. R. L., Richards, T. G., and Short, A. H. (1966). Factors determining the maximal rate of organic anion secretion by the liver and further evidence on the hepatic site of action of the hormone secretin. *J. Physiol. (Lond.)* **186:**424–438.

Paterson, J. Y. F., and Harrisonn, F. A. (1972). The splanchnic and hepatic uptake of cortisol in conscious and anesthetized sheep. *J. Endocrinol.* **55:**335–350.

Pries, J. M., Staples, A. B., and Hansen, R. F. (1981). The effect of hepatic blood flow on taurocholate extraction by the isolated perfused rat liver. *J. Lab. Clin. Med.* **97:**412–417.

Raymond, G. D. (1966). *Gastroenterology* **50:**862–863.

Reichen, J., and Paumgartner, G. (1976). Uptake of bile acids by perfused rat liver. *Am. J. Physiol.* **231:**734–742.

Reichen, J., and Paumgartner, G. (1980). Excretory function of the liver. In *Liver and Biliary Tract Physiology I*, Int. Rev. Physiol., Vol. 21, Javitt, N. B., ed., Baltimore: University Park Press, pp. 103–150.

Roberts, R. J., Klaassen, C. D., and Plaa, G. L. (1967). Maximum biliary excretion of bilirubin and sulfobromophthalein during anesthesia-induced alteration of rectal temperature. *Proc. Soc. Exp. Biol. Med.* **125:**313–316.

Roberts, R. J., and Plaa, G. L. (1967). Alterations of the plasma disappearance and biliary excretion patterns of exogenously administered bilirubin by α-napthylisothiocyanate. *J. Pharmacol. Exp. Ther.* **155:**330–336.

Roberts, R. J., and Plaa, G. L. (1967). Effect of phenobarbital on the excretion of an exogenous bilirubin load. *Biochem. Pharmacol.* **16:**827–835.

Roberts, R. J., Shriver, S. L., and Plaa, G. L. (1968). Effect of norethandrolone on the biliary excretion of bilirubin in the mouse and rat. *Biochem. Pharmacol.* **17:**1261–1268.

Rowland, M., and Tucker, G. (1980). Scientific commentary. Symbols in pharmacokinetics. *J. Pharmacokin. Biopharm.* **8:**497–507.

Ryan, J., Cooper, M., Loberg, M., Harvey, E., Sikorski, S.: Technetium-99m-labeled *N*-(2,6-dimethylphenylcarbamoylmethyl)iminodiacetic acid (Tc-99m HIDA): A new radiopharmaceutical for hepatobiliary imaging studies. *J. Nucl. Med.* **18:**995–1002.

Schanker, L. S. (1968). Secretion of organic compounds into bile. In *Handbook of Physiology*, Vol. 5, Code, C. F., and Heidel, W., eds. Washington, D.C.: American Physiology Society, pp. 2433–2450.

Schwenk, M. (1980). Transport systems of isolated hepatocytes. Studies on the transport of biliary compounds. *Arch. Toxicol.* **44:**113–126.

Smith, R. L. (1973). Species variations in biliary excretion. In *The Excretory Function of Bile*. London: Chapman and Hall, pp. 76–93.

Smith, R. B., Coupal, J., DeLand, F. H., and Triplett, J. W. (1979). Pharmacokinetics of hepatobiliary imaging agents in rats. *J. Nucl. Med.* **20:**45–49.

Upson, D. W., Gronwall, R. R., and Cornelius, C. E. (1970). Maximal hepatic excretion of bilirubin in sheep. *Proc. Soc. Exp. Biol. Med.* **134:**9–12.

Wagner, J. G. (1976). Scientific commentary: Linear pharmacokinetic equations allowing direct calculation of many needed pharmacokinetic parameters from the coefficients and exponents of polyexponential equations which have been fitted to the data. *J. Pharmacokinet. Biopharm.* **4:**443–467.

Weinbren, K., and Billing, B. H. (1956). Hepatic clearance of bilirubin as an index of cellular function in the regenerating rat liver. *Br. J. Exp. Pathol.* **37:**199–204.

Whelan, G., and Combes, B. (1975). Phenobarbital enhanced biliary excretion of administered unconjugated and conjugated sulfobromophthalein in the rat. *Biochem. Pharmacol.* **24:**1283–1286.

Wilkinson, G. R., and Shand, D. G. (1975). A physiological approach to hepatic drug clearance, clinical pharmacology and therapeutics. *Clin. Pharmacol. Ther.* **18:**377–390.

Wistow, B. W., Subramanian, G., VanHeertum, R. L., Hendersen, R. W., Gagne, G. M., Hall, R. C., and McAfee, J. G. (1977). An evaluation of ^{99m}Tc-labeled hepatobiliary agents. *J. Nucl. Med.* **18:**455–461.

Zimmerman, H. J. Hepatotoxicity: *The Adverse Effects of Drugs and Other Chemicals on the Liver.* New York: Appleton-Century-Crofts, pp. 180–185, 259–276.

CHAPTER 7

Experimental Models and Evaluation of Animal Data for Renal Radiodiagnostic Agents

JOHN G. McAFEE and GOPAL SUBRAMANIAN

INTRODUCTION

After a new gamma-emitting radiodiagnostic agent is synthesized and its chemical structure defined, serial camera imaging following administration to experimental animals can quickly determine its major organ localization. Blood and plasma disappearance curves and urinary excretion at different time intervals are also measured. For renal agents, with negligible extrarenal concentration or excretion, a double exponential analysis of the plasma disappearance curve may be used to quantitate renal clearance, in dogs (Blaufox et al., 1963) or even in small animals like the rat (Blaufox et al., 1970). If a new agent appears promising for eventual human use, more detailed concentration values are required for various organs at different time intervals. These are generally obtained by direct tissue radioassay following the sacrifice of a series of animals. It is frequently rewarding to compare (by dual channel pulse height analysis) the distribution of a new agent with that of an older one with similar organ localization, simultaneously injected and labeled with a radionuclide of different gamma energy. These tissue concentration data from animals are essential to calculate preliminary radiation dose estimates, before administration to humans is considered.

EVALUATION OF ANIMAL BIODISTRIBUTION DATA

For analysis of data from different species or for animals of different size within the same species, the concentration of radionuclide may be expressed as percent administered activity per total organ (Blau, 1975). The measurement of activity in total organs, however, is difficult in large animals and for some tissues such as the skeleton, musculature, marrow, gastrointestinal tract, and blood. Tissue sample data are often expressed as percent administered activity per gram; however, this is unsatisfactory for comparing data from animals of different mass. For example, suppose an agent "X" concentrates uniformly in all tissues without preferential uptake or excretion and its distribution is evaluated in two rats weighing 100 and 200 g. If the concentration of X is calculated in percent administered activity per gram of tissue, this value is 1 in one animal and 0.5 in the other. These values give a false impression of variation in concentration where none exists. Hence various expressions similar in concept have been described (Kenney et al., 1941; Elgee et al., 1955; Long et al., 1963; Oldendorf, 1974; Woodard et al., 1975; Kirschner et al., 1975) that normalize tissue concentration data according to body weight (summarized in Fig. 7-1). When one employs these methods of expression, variations in weight between animals in a group under study do not influence the results. Moreover, these expressions enable interspecies comparisons of concentration data.

Extrapolation of distribution data of radionuclides from animals to man is difficult and sometimes inaccurate (Stara et al., 1971). The risk of extrapolation errors is diminished when several mammalian species are studied. Species differences may be arbitrarily divided into two types: (1) unpredictable species anomalies often due to differences in metabolic pathways or protein binding and unrelated to animal size and (2) predictable differences related to the circulation time and body size. There appears to be no general solution

References	*Expression*	*"Uniformity" Value+*
Kenney et al. (1941)	Differential absorption $= \dfrac{\mu\text{Ci found/kg. tissue}}{\mu\text{Ci admin./kg. body wt}}$	1
Elgee and Williams (1955)	Concentration [T] / [B] = % total dose/gm × gm body wt /100	1
Long et al. (1963)	% Admin. activity/1% body wt $= \dfrac{\text{cpm tissue} \times \text{gm body wt}}{\text{cpm admin.} \times \text{gm tissue}}$	1
Oldendorf (1974)	% Mean body concentration $= \dfrac{\text{cpm/gm tissue} \times 100}{\text{injected cpm/gm total animal}}$	100
Woodard et al. (1975)	Relative concentration RC $= \dfrac{\mu\text{Ci found/gm tissue}}{\mu\text{Ci administered/gm body wt}}$	1
Kirschner et al. (1975)	% kg dose/gm $= \dfrac{\mu\text{Ci in organ/gm} \times 100}{\mu\text{Ci (dose)/kg body wt}}$	0.1

FIG. 7-1. Methods of expressing tissue concentration data. +"Uniformity" value corresponds to no preferential uptake or elimination of the tracer from the tissue being examined.

TABLE 7-1. Parameters Derived from Body Weight in Different Species

Species	*Body Weight (gm)*	K *Value*[a]	*Surface Area* $(M^2)^b$	$W^{1/3}$	*Man-min Equivalent*[c] *(min)*	*Equivalent Time Factor*[d]
Man	70,000	10.6	1.80	41.2	1	1
Dog	10,000	10.1	0.469	21.5	0.52	1.92
Monkey	4,000	11.8	0.297	15.9	0.38	2.63
Rabbit	3,000	6.0	0.125	14.4	0.35	2.86
Rat	250	9.0	0.0357	6.3	0.15	6.67
Mouse	20	9.0	0.00663	2.7	0.066	15.1

[a] *K* values from Spector (1956).
[b] $M^2 = KW^{2/3} \times 10^{-4}$.
[c] Man-min equivalent = $W^{1/3}$ animal/$W^{1/3}$ man.
[d] Reciprocal of c.

for the first type, other than the study of several mammalian species, but if only the second type exists, the behavior of stable drugs or radiodiagnostic agents in man can be reasonably predicted from animal data.

In many comparative studies of stable drugs (Freireich et al., 1966; Dedrick et al., 1970), plasma disappearance is most rapid in small rodents, intermediate for the dog and monkey, and slowest in man. Relative to man, the time scale appears to be compressed in smaller animals. They have a shorter life span and higher metabolic, cardiac, and respiratory rates, so that each animal species lives approximately the same number of cardiac and respiratory cycles. Because the mean circulation time (blood volume/cardiac output) is shorter in smaller animals, renal agents, in particular, are excreted faster when more cycles of the total plasma volume pass through the kidneys per minute than in man, even though their renal extraction efficiencies are similar. In a comparative study of clearance of chemotherapeutic agents, Dedrick et al. (1970) found that these differences largely disappear if the time scale for each species is altered as a power function of the body weight (W). They empirically chose a value of $W^{0.25}$ to calculate "man equivalent time"; for many radiopharmaceuticals, however, we (McAfee et al., 1981a) obtained a better "fit" with blood or plasma disappearance and total body retention using $W^{1/3}$. Thus for each species the "equivalent-time" factor is calculated as $W^{1/3}$ animal/$W^{1/3}$ man (see Table 7-1). Hence to expand the animal time base to "man equivalent time" one multiplies clock time bv the reciprocal of $W^{1/3}$ man/$W^{1/3}$ animal in column 7 of Table 7-1.

COMPARISON OF THE BIODISTRIBUTION OF RENAL AGENTS IN DIFFERENT SPECIES

Figures 7-2 to 7-5 show examples of blood clearance and body retention for common renal agents, altering the time scale for each species by multiplying real time by the "equivalent-time factor" in comparison with unaltered time for human data. Often the experimental animal data adequately predict the human values within ±1 standard deviation.

Figure 7-2A shows the real-time blood disappearance curves for [^{99m}Tc]DTPA, a nonmetabolized agent with an extracellular distribution cleared by glomerular filtration. The blood clearance typically is faster in the smaller species. On altering the time scale (Fig. 7-2B) the blood levels are surprisingly close in different species. It will be noted that slight errors in recording the real time of sampled measurements in small rodents magnify in

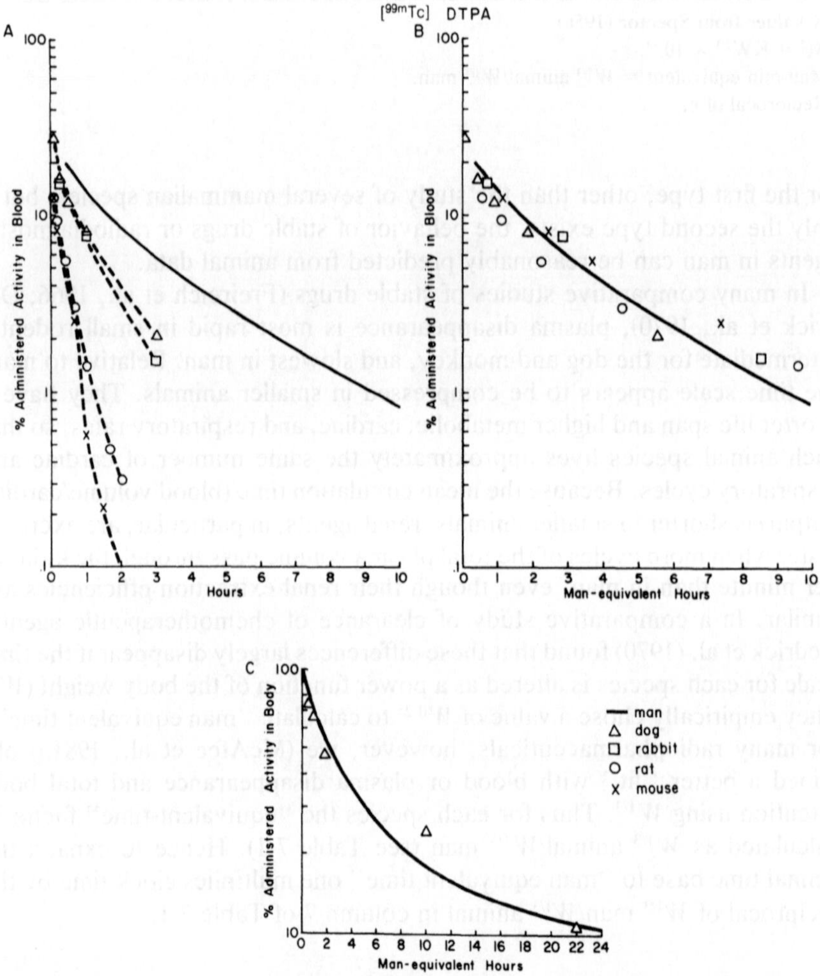

FIG. 7-2. ^{99m}Tc-DTPA. (A) Blood clearance in different species plotted conventionally in real time. (B) Blood clearances plotted in "man-equivalent" time = real time × $W^{1/3}$ man/$W^{1/3}$ animal. (C) Body retentions plotted in "man-equivalent" time. (All data in figures corrected for radioactive decay.)

the expanded time scale. The normalized total body retention for the dog reasonably predicts human retention of this agent up to 24 hr (Fig. 2C). Thus the apparent species differences for this agent are entirely due to body size.

[^{99m}Tc]pertechnetate has a predominantly extracellular distribution except for the thyroid, gastric mucosa, and salivary glands. It is actively reabsorbed by the renal tubules, so that its renal clearance is only about 14% of the inulin clearance, and more is excreted by the gastrointestinal tract than by the kidneys. Its real-time blood disappearance in rodents is again faster than that in man (Fig. 7-3A), but this difference tends to disappear with the time-equivalent scale (Fig. 7-3B). The normalized total body retention in the dog and man are similar up to one week (Fig. 7-3C).

The rather meager blood clearance data for the renal agents [^{99m}Tc]dimercaptosuccinate (DMS) and glucoheptonate (GHA) (Arnold et al., 1975) plotted on a time-equivalent scale are shown in Fig. 7-4. There is good agreement between man and rat for both agents and between man and rabbit for DMS, but the single values in the dog are considerably lower.

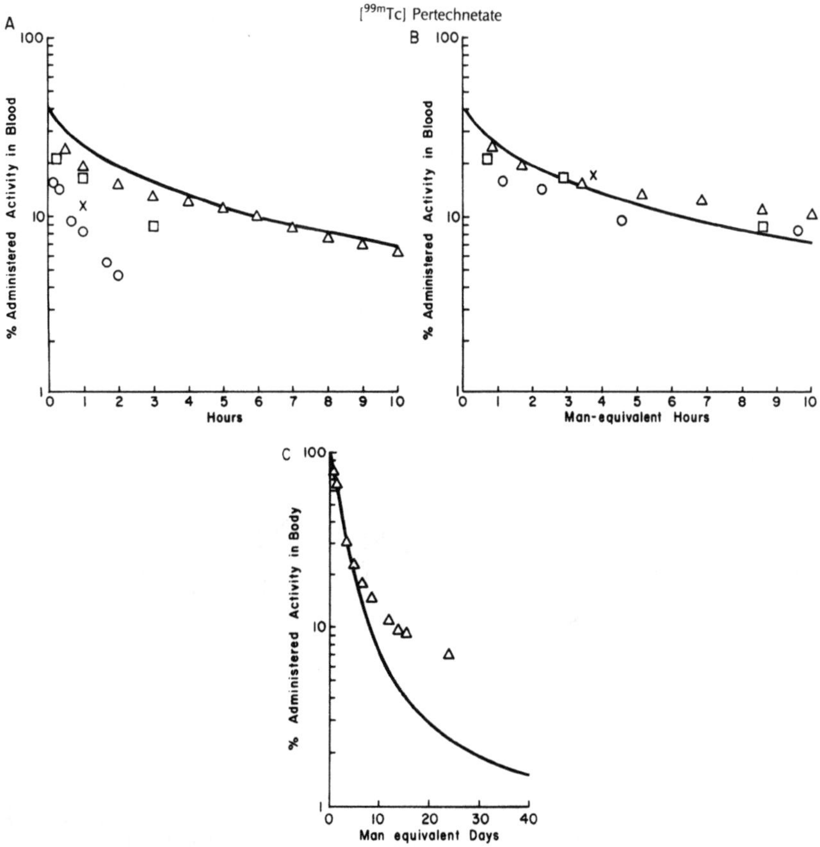

FIG. 7-3. [^{99m}Tc]pertechnetate blood clearances (A) plotted in real time, (B) plotted in man-equivalent time. (C) Body retentions of [^{95m}Tc]pertechnetate.

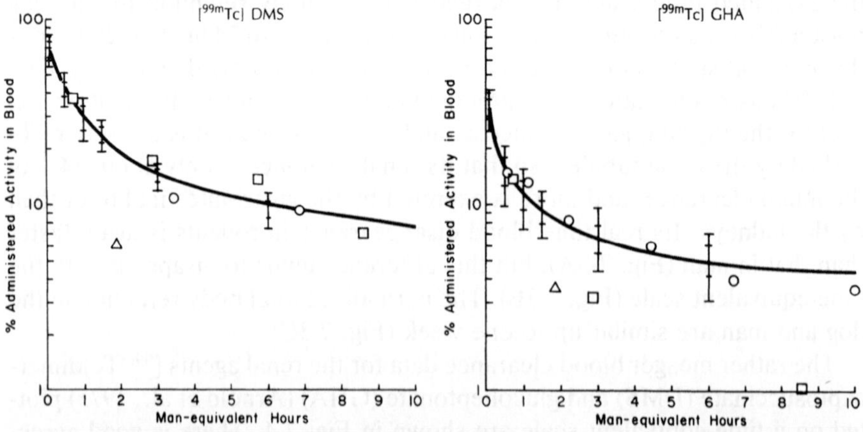

FIG. 7-4. Blood clerances of [^{99m}Tc]DMS and GHA plotted in man-equivalent time.

The method of expanding the time scale of animals based on $W^{1/3}$ somewhat overcorrects the blood clearance of [^{131}I]orthoiodohippurate (Hippuran) in smaller species (Fig. 7-5A). This agent has been used widely to estimate the effective renal plasma flow (ERPF) because its plasma protein binding is minimal and its renal extraction efficiency, by a combination of glomerular filtration and tubular secretion, is extremely high. Previous inter-

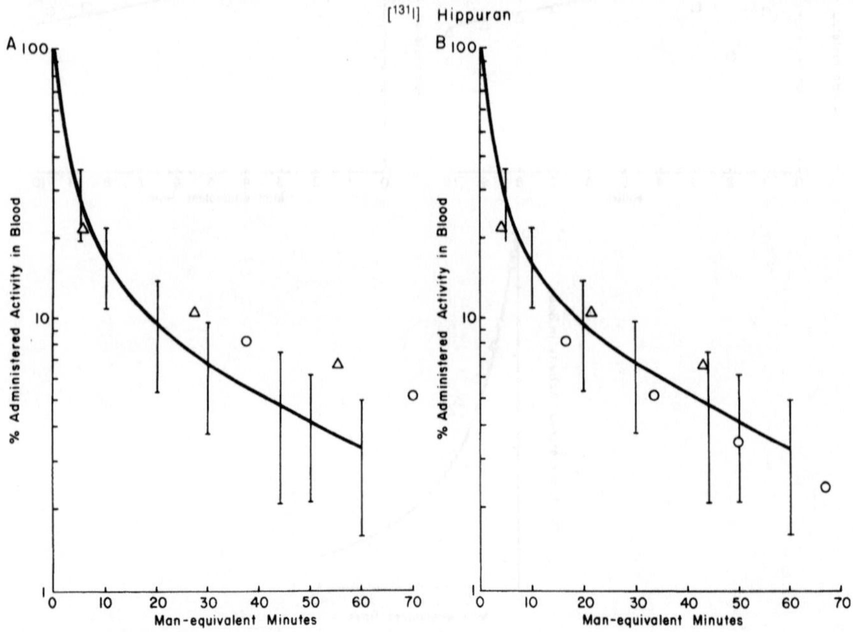

FIG. 7-5. Blood clearance of [^{131}I]hippuran (A) time scale based on $W^{1/3}$, (B) time scale based on $W^{.2}$.

species comparisons (Adolph, 1949) have shown the clearance of para-aminohippurate (PAH) and renal blood flow correlate with $W^{0.8}$. Hence normalization of the time scale for [^{131}I]hippuran based on a factor of $W^{0.2}$ resulted in a better fit of blood clearance data in different species (Fig. 5B).

For a long time, biologists have related many physiological functions in different species to power functions of body weight (McMahon 1973), and many are related to body surface area (from $W^{2/3}$), the so-called Voit's surface area law. Certain functions in man are still normalized to surface area rather than body weight, such as cardiac output, renal blood flow, glomerular filtration rate, chemotherapeutic drug doses, pediatric doses, and drug toxicity levels. Many functions, however, are proportional to $W^{0.75}$ (Kleiber's rule). Adolph (1949) computed power functions of many parameters in various species and found that no single value predominated. As examples, the exponent of body weight for water intake was 0.68; creatinine clearance, 0.69; hippurate clearance, 0.8; urinary output, 0.82; and inulin clearance, 0.77. Such biological frequency functions as heart rate, respiratory rate, or intestinal contraction rate have been correlated negatively with body weight, usually by $W^{-1/3}$ or $W^{-0.25}$ (McMahon 1973). From the experience of biologists it can be concluded that the prediction of physiological functions or drug distributions from one species to another based on power functions of body weight are only approximate.

From the limited biodistribution data on radiopharmaceuticals currently available for interspecies comparisons, our normalization procedure, based on $W^{1/3}$, appears to minimize species differences for extracellular agents excreted largely by glomerular filtration. However, the procedure is ineffective for agents localizing predominantly intracellularly, such as ^{201}Tl-Cl,[^{75}Se]selenomethionine or [^{131}I]iodocholesterol. Dedrick et al. (1970) anticipated failure of this approach for lipid-soluble drugs, those strongly bound to plasma proteins or tissue cells, or those excreted by nonlinear processes.

The organ distribution of the common renal agents in laboratory animals and man are compared in Table 7-2. These data are incomplete, particularly for the oldest agent, [^{131}I]hippuran. Considerable interspecies variations are evident. The data for the ^{99m}Tc renal agents again indicate higher blood levels and slower urinary excretion in man than in smaller animals. A high anomalous renal uptake of dimercaptosuccinate (DMS) is noted in the rat and mouse.

CHOICE OF ANIMAL MODELS

A comparison of certain fundamental renal parameters in different laboratory species is given in Table 7-3. The mouse is inexpensive and easy to handle in the laboratory. However, it is seldom used by renologists, so that relatively little is known about its renal physiology or spontaneous renal

TABLE 7-2. Distribution of Common Renal Agents in Animals and Man—Percent Administered Activity per Organ

[^{131}I]hippuran

	Mouse	Rat		Rabbit	Dog		Man	
		3 min	1 hr	1 hr	3 min	1 hr	3 min	1 hr
Blood	—	48	0.58	1.1	22	3.9	31	3.3
Two kidneys	—	36	0.94	—	20	1.5	—	—
Liver	—	12	0.63	—	6.7	1.5	—	—
Muscle	—	—	2.0	—	17	7.6	—	—
Urine	—	—	71	93	6.2	71	0	78

[^{99m}Tc](Sn) DTPA

	Mouse	Rat	Rabbit	Dog		Man	
		1 hr	1 hr	3 min	1 hr	3 min	1 hr
Blood	—	2.3	8.6	28	8	72	15
Two kidneys	—	2.2	3.4	8.8	2.0	14	5.2
Liver	—	.41	1.9	3.5	1.2	3.5	2.3
Muscle	—	—	5.4	22	9.1	—	—
Urine	—	—	53	3.3	53	0	25

[^{99m}Tc]glucoheptonate (GHA)

	Mouse	Rat	Rabbit	Dog	Man	
	1 hr	6 hr	1 hr	1 hr	1 hr	3 hr
Blood	—	0.7	4.0	1.8	11	6.6
Two kidneys	4.6	23	13	9.6	9.2	11.8
Liver	2.2	1.0	1.5	1.7	—	—
Muscle	—	1.4	3.2	6.0	—	—
Urine	—	—	61	—	38	55

[^{99m}Tc]dimercaptosuccinate (DMS)

	Mouse		Rat	Rabbit	Dog	Man	
	1 hr	3 hr	4 hr	1 hr	1 hr	1 hr	3 hr
Blood	27	5.8	9.3	17	3.2	30	14
Two kidneys	40	22	52	20	17	25	37
Liver	9.4	5.9	4.6	3.2	1.8	—	—
Muscle	—	7.3	3.7	9.8	8.7	—	—
Urine	16	—	15	15	—	11	19

diseases. Its single nephron glomerular filtration rate (SNGFR) is exceptionally low among species. Moreover, it is too small for renal imaging.

The rat is the favorite model of renologists, because so much is known about its renal anatomy and physiology, and many induced renal lesions have been studied (to be discussed later) that mimic various human renal diseases. Because they are relatively cheap, they may be used in large numbers. They are sufficiently large for imaging with gamma cameras, particularly with pinhole collimators.

TABLE 7-3. Comparison of Renal Structure and Function in Different Species[a]

	Body Wt (kg)	*Wt 2 Kidneys (g)*	*% Body Wt 2 Kidneys*	*No. of Glomeruli, 2 Kidneys*	*GFR (ml/min)*	*GFR (ml/min M^2)*	*SNGFR[b] (nl/min)*	*ERPF (ml/min)*	*ERPF (ml/min M^2)*	*FF[c]*	*Urine Output (ml/day)*
Mouse	0.03	0.4	1.33	2.4×10^4	0.2	23	9	—	—	—	—
Rat	0.26	2.0	0.77	6×10^4	2.0	55	32	7.1	196	28	11
Rabbit	3.0	24	0.81	4×10^5	9.2	52	23	54	300	17	270
Cat	3.0	30	1.0	3.6×10^5	6.0	29	15	29	138	21	78
Dog	10.0	70	0.7	8×10^5	43	92	50	134	287	32	310
Human	70.0	300	0.43	2×10^6	130	75	65	650	376	20	1500

[a] Data from various sources.
[b] SNGFR = single nephron glomerular filtration rate.
[c] FF = filtration fraction (GFR ÷ ERPF) × 100

One disadvantage of this species for renal studies is spontaneous glomerular sclerosis (chronic nephropathy) in aging rats (Bolton et al., 1976). Most strains of albino laboratory rats develop this degenerative kidney disease of unknown cause, increasing in severity and incidence with advancing age. Its onset is characterized by proteinuria, mostly of albumin, and later, hypercholesterolemia. It is more common in males than females. Histologically, the early disease is predominantly glomerular, with thickening of the basement membrane and mild focal hyperplasia associated with mesangial deposits of IgM. The lesions only superficially resemble focal glomerular sclerosis in man, and there is little evidence to link the rat lesions to immune mechanisms. Fischer (F-344) rats are relatively susceptible, and Lewis rats are less susceptible than most strains. Random-bred male Sprague–Dawley rats have an incidence of proteinuria of 25% at 12 weeks of age, 55% at 1 year, and 95% at 2 years (Bolton et al., 1976). Older rats have a high incidence of spontaneous myocardial disease also.

For studies of urinary excretion of radioactive agents, we prefer male Sprague–Dawley rats. For 2-hour experiments with anesthetized animals, the male urethra can be easily clamped for collection of urine from the bladder. However, for longer experiments, metabolic cages have to be used for urinary collections. For control experiments we use animals about 8 weeks old and with a body weight of about 200 g. To avoid the problem of including rats with spontaneous nephropathy, we reject animals weighing more than 300 g (over 8–10 weeks of age).

It is tempting to use rabbits for renal imaging because of their large size and ease of handling, and they are less expensive than dogs. They have been employed extensively in the study of immune complex glomerulopathy (serum sickness disease) induced by the injection of heterologous sera (Lewis et al., 1971). The rabbit is the only animal other than the flounder in which individual tubules can be dissected with the basement membranes intact. Nevertheless, the rabbit exhibits many peculiarities in renal function that make it an unsatisfactory model for the measurement of renal clearance of many substances (Brewer, 1977). Its urinary output is very high relative to body weight or surface area compared to other mammals (Table 7-3). It is the only mammal in which glomerular activity remains intermittent in the adult. The number of active glomeruli are increased by water hydration and maximally increased with saline, resulting in tremendous increases in urine flow. This interdependence of urine flow, GFR, and ERPF is not seen in adults of other species (Dicker et al., 1935).

Splanchnic neurogenic stimulation in rabbits causes a marked drop in renal blood flow because of intense vasoconstriction of the arteries of the peripheral cortex with ischemia, while the juxtamedullary arteries remain open to perfuse the medulla (Trueta phenomenon). This intense reaction, leading to oliguria or anuria, is absent or mild in other species. This reflex reaction in rabbits may be elicited by trivial stress, resulting in a sudden marked reduction in renal blood flow (Brewer, 1977). On the other hand, renal blood flow and GFR change much less than in other species with alterations in blood pressure. Differences in renal metabolism between the rabbit and other species have been uncovered, particularly the secretion of ammonia and transport of uric acid.

Certain experiments are most conveniently carried out in dogs, particularly when open surgery is required. For continuous infusions with urine collection, the bladder or ureters may be catheterized. For repeated experiments of this type, some workers perform a bladder-splitting procedure for the separate collection of urine from each kidney. The dog is particularly useful for renovascular lesions. The renal arteries and veins may be selectively catheterized by the percutaneous Seldinger technique identical to the clinical method. The renal extraction efficiency of radiodiagnostic agents is conveniently measured by simultaneous catheterization of a renal artery and vein (McAfee et al., 1981). A model of unilateral acute renal artery stenosis is produced without open surgery by percutaneous insertion of a Swan–Ganz catheter into a main renal artery and inflating the balloon stepwise to the desired reduction in renal arterial pressure (Korobkin et al., 1974). The pressure is measured through the end-hole catheter. To produce "chronic" renal artery stenosis, a circumferential balloon constrictor is implanted surgically. Periarterial blood flow transducers are often placed around both renal arteries for comparison with radioisotopic blood flow measurements. Intrarenal cortical and medullary blood flow may be measured by analysis of the washout curve of ^{133}Xe in saline injected into a renal artery. To produce renal infarction, a renal artery or one of its branches may be ligated. An infarct may be produced without surgery by releasing a metallic bead from the tip of a catheter percutaneously inserted into a renal artery. For temporary renal ischemia, a circumferential balloon constrictor around a renal artery may be temporarily inflated. Most experiments on autologous or homologous renal transplants have been conducted in dogs.

Many species have been used as a model of obstructive uropathy by ureteral ligation. Producing incomplete obstruction, however, is more difficult. A technique for controlling and monitoring pressure in the renal pelvis in dogs was developed by Jones et al. (1963). The left ureter is transected and two #5 ureteral catheters are passed retrograde into the renal pelvis and the ureter then ligated. One catheter is connected to a pressure transducer and the other is connected by polyethylene tubing to a flask of saline. The height of the flask controls the hydrostatic pressure in the renal pelvis. Raising or lowering the flask to known levels mimics different degrees of obstruction, or intermittent obstruction.

A model of pyelonephritis in dogs was developed by Bricker et al. (1958). One kidney is surgically exposed through the flank and the upper ureter is clamped for 30 min. After massaging the kidney to produce hyperemia, 100–150 needle punctures of the kidney to a depth of 6 mm are made. Then 1 ml of bacterial culture is injected intravenously and the ureteral clamp is removed.

The newborn piglet is the preferred model for reflux nephropathy (Hodson et al., 1975) because the pig possesses a multipyramidal multipapillate kidney like the human organ. Reflux is created by resection of the roof of the intravesical ureter and anterior incision of the ureterovesical junction (Ransley and Risdon, 1978). In addition, Hodson et al. (1975) used a silver constrictor around the proximal urethra just below the internal sphincter. Ascending infection may be induced by instilling porcine coliform bacillus culture into the bladder (Ransley et al., 1978). Because the pelvocalyceal anatomy of the pig is similar to that of man, Hodson considers the adult pig the best model for obstructive uropathy, produced by temporary or permanent ligation of one or both ureters.

The egg-laying chicken possesses unique anatomical features suitable for certain renal experiments (Rennick et al., 1956). The two ureters of hens have separate openings into the urodaeum so that they can be easily cannulated externally to obtain urine collections separately from each kidney. Also, they have an unusual renal "portal" venous system from the lower extremity veins. When an agent is injected into one saphenous vein, it reaches the renal parenchyma through the portal circulation before it reaches the inferior vena cava. The difference in urinary excretion between the injected and opposite side may be used as an index of tubular excretion. Substances not excreted by the tubular cells are excreted equally by the two kidneys.

MODELS OF HUMAN RENAL DISEASES IN THE RAT

To induce renal failure, the so-called five-sixth nephrectomy has the disadvantage of a high mortality, and bilateral ureteral ligation is fatal within 3 days (Giacomina et al., 1981). Among nephrologists, uranyl nitrate is a popular agent both for inducing renal failure or lesser degrees of renal tubular damage. Five milligrams per kilogram intravenously causes some impairment in renal function at one day, becoming marked by 6 days. This dose causes some decrease in glomerular permeability, but there is little or no effect on liver function. Histologically, necrosis of the distal straight portion of the proximal renal tubules at the corticomedullary junction is apparent by 4 days. The mortality from this dose of uranyl nitrate is very low.

Cis-Platinum consistently causes tubular damage in male Sprague–Dawley rats. In our experience, five daily intravenous doses of 0.36, 0.71, and 1.4 mg/kg produce acute tubular necrosis of slight, moderate, and marked severity, respectively, within 7 days after the last dose. In the susceptible Fisher-344 rat, a single intraperitoneal dose of 6 mg/kg induces pathological changes at 3 days and widespread tubular necrosis at 5 days (Dobyan et al., 1980). The histological and functional changes of gentamicin nephrotoxicity also have been thoroughly studied in rats (Black et al., 1964). In our hands, about 50% of rats given 40 mg/kg intravenously daily for 5 days will develop some tubular necrosis about 1 week later. A majority will develop more severe tubular necrosis or die when the daily dose is increased to 90 mg/kg.

Biochemical alterations are apparent within hours after injection of high doses (Williams et al., 1981). With 100 mg/kg daily subcutaneously for 3–5 days, the GFR is depressed, but there is an anomalous increase in PAH clearance above control values (Cohen et al., 1975). Many other agents have been employed intravenously to create acute tubular necrosis, including cadmium, mercury or arsenic salts, sodium ferrocyanide, cyanide, tartrate, trimethadone, iodoacetate, phlorizin, *N*-ethyl maleimide, methemoglobin, and angiotensin II.

Turning to glomerular lesions, there is a model of "minimal change" nephrotic syndrome (proteinuria edema, lipidemia, and hypoproteinemia) common in children and usually responding to steroids. This condition is mimicked by administering puromycin aminonucleoside (6-dimethyl aminopurine 3-amino-D-ribose) usually to male Sprague–Dawley rats. In the initial technique (Frenk et al., 1955), subcutaneous injections of 1.5 mg/100 g in 2% saline are administered daily for 10 days. By proper selection of the dose, there may be eventual recovery. In a simpler technique—single intravenous injection of 5–10 mg/100 g—proteinuria usually appears about 5 days, maximal by about 10 days, with only partial recovery by about 3 weeks (Lannigan, 1963). As a rule, the kidneys appear normal by light microscopy, but, by EM, fusion of the epithelial foot processes is demonstrated (Ryan et al., 1975). A technique has been developed for producing this lesion in only one rat kidney by unilateral renal arterial perfusion and transient left renal vein drainage (Hoyer et al., 1975). This technique may be used also in dogs. Protamine injected intravenously (3 mg/100 g) induces glomerular changes for several hours, simulating the effects of puromycin aminonucleoside with fusion of renal epithelial foot processes (Sharon et al., 1977). These changes can be promptly reversed by heparin infusion (Seiler et al., 1977).

Immune-complex-mediated glomerulopathies may be induced in many species by injection of "foreign" serum proteins. Injections of preformed soluble antigen–antibody complexes that fix complement produce lesions similar to that of human serum sickness or poststreptococcal glomerulopathies (Lewis et al., 1971). Another rat experimental model, nephrotoxic serum nephritis of Masugi, is produced by immunizing rabbits with rat kidney tissue and injecting the antiserum into rats. In this lesion the antigen is the rat glomerular basement membrane and probably has no exact counterpart in human disease. A lesion identical to human membranous nephropathy is produced in rats by immunization with a proximal tubular brush border antigen (Heymann nephritis model) (Salant et al., 1980). The rat is immunized by intraperitoneal injection of homologous renal tissue in complete Freund's adjuvant. The same lesion is produced more rapidly by a single injection of heterologous brush border antibody (passive Heymann nephritis model). These models of immune-complex-associated glomerulopathies have not been used as yet in nuclear medicine but may become important if renal agents with more specific tissue localization are developed.

The rat is frequently used in experimental hypertension. This condition occurs spontaneously in many rat colonies, increasing in incidence with age. One strain was produced by selective breeding through 18 generations in Japan by Okamoto and Aoki (1963)—the spontaneous hypertensive (SHR) rat. This strain develops hypertension independent of salt intake and is now widely used. Two other strains were developed at Brookhaven National Laboratories by Dahl (1964) by selective breeding of Sprague–Dawley rats over three generations and feeding the weanlings a high-salt diet. One group, the Dahl Sensitive (S) strain develops hypertension on an 8% salt diet and the other Dahl Resistant (R) strain does not. These strains are useful for studying the role of sodium in hypertension. However, the equivalence of rat to human hypertension has been questioned, largely because of marked differences in intrarenal prostaglandin metabolism (McGiff et al., 1981).

We have produced renal infarction in the rat by applying a microvascular clamp (0.5–1.1 mm vessel size) to a renal artery branch with 10 g pressure under a dissecting microscope. Clamping for 30 min or more results in irreversible necrosis; recoverable ischemia is produced by clamping from 10–20 min.

Nephrocalcinosis can be induced in rats with large doses of vitamin D_3 (Potokar et al., 1978). Multipe calcium oxalate renal calculi can be produced in about 80% of young male rats on a vitamin B_6 (pyridoxine)-deficient diet for 6–8 weeks (Andrus et al., 1960).

To develop pyelonephritis in the rat, *E. coli* are directly inoculated at three sites into the surgically exposed kidney through a glass microcapillary (Miller et al., 1973). Renal papillary necrosis may be produced in about 3 days by a single intravenous injection of ethyleneimine (10 mg/kg) (Sherwood et al., 1971).

The principal features of the Fanconi syndrome are aminoaciduria, glycosuria, phosphaturia, and acidosis. This human congenital renal tubular defect is mimicked by the injection of neutralized maleic acid intravenously (40 mg/kg) (Berliner et al., 1950) or subcutaneously (40 mg/kg) (Rosenberg et al., 1964) in rats or dogs. The abnormal urinary excretion appears in 1 or 2 hours and persists for several hours, often without changes in GFR.

CONCLUSIONS

In the evaluation of new radiotracers or new diagnostic applications of established radiopharmaceuticals, predicting their biological distribution in man from animal data is often rewarding but imprecise and sometimes impossible. The risk of interspecies extrapolation errors generally diminishes when several mammalian species are studied. For renal agents with predominantly extracellular distribution, excreted by glomerular filtration or renal tubular secretion, interspecies variations may be minimized by normalization with power functions of body weight. Animal models and models of human diseases for testing radioactive agents must be selected with care from the accumulated knowledge in biology, pathology, and pharmacology, so that experimental results will be relevant to their potential clinical application. From the experience in pharmacology (Mellett, 1969), no species of laboratory animal, primate or otherwise, will consistently excrete or metabolize drugs at the same rate or in the same fashion as man. Similar limitations apparently apply to radiopharmaceuticals.

ACKNOWLEDGMENT

This investigation was supported in part by PHS Grant No. CA 32848 awarded by The National Cancer Institute, DHHS.

REFERENCES

Adolph, E. F. (1949). Quantitative relations in the physiological constitution of mammals. *Science* **109:**579–585.

Andrus, S. B., Gershoff, S. N., Faragalla, F. F., and Prier, E. L. (1960). Production of calcium oxalate renal calculi in vitamin B_6-deficient rats. *Lab. Invest.* **9:**7–27.

Arnold, R. W., Subramanian, G., McAfee, J. G., Blair, R. J., and Thomas, F. D. (1975). Comparison of ^{99m}Tc complexes for renal imaging. *J. Nucl. Med.* **16:**357–367.

Berliner, R. W., Kennedy, T. J., and Hilton, J. G. (1950). Effect of maleic acid on renal function. *Proc. Soc. Exp. Biol. Med.* **75:**791–794.

Black, J., Calesnick, B., Williams, D., and Weinstein, M. J. (1964). Pharmacology of gentamicin, a new broad-spectrum antibiotic. *Antimicrobiol. Agents Chemother.* **1963:**138–147.

Blaufox, M. D., Orvis, A. L., and Owen, C. A., Jr. (1963). Compartmental analysis of the radiorenogram and distribution of Hippuran I^{131} in dogs. *Am. J. Physiol.* **204:**1059–1064.

Blaufox, M. D., and Cohen, A. (1970). Single-injection clearances of iothalamate—^{131}I in the rat. *Am. J. Physiol.* **218:**542–544.

Blau, M. (1975). Radiation dosimetry of ^{131}I-19-iodocholesterol: The pitfalls of using tissue concentration data. *J. Nucl. Med.* **16:**247–248.

Bolton, W. K., Benton, F. R., Maclay, J. G., and Sturgill, B. C. (1976). Spontaneous glomerular sclerosis in aging Sprague–Dawley rats. *Am. J. Pathol.* **85:**277–300.

Brewer, N. R. (1977). Some oddities of the rabbit kidney. *Synapse* (American Society of Laboratory Animal Practioners) **10:**19–22.

Bricker, N. S., Dewey, R. R., and Lubowitz, H. (1958). Studies in experimental pyelonephritis: Simultaneous and serial investigation of a pyelonephritic and intact kidney in the same animal. *Clin. Res.* **6:**292–293.

Cohen, L., Lapkin, R., and Kaloyanides, G. J. (1975). Effect of gentamicin on renal function in the rat. *J. Pharmacol. Exp. Ther.* **193:**264–273.

Dahl, L. K., and Schackow, E. (1964). Effects of chronic excess salt ingestion: Experimental hypertension in the rat. *Can. Med. Assoc. J.* **90:**155–160.

Dedrick, R. L., Bischoff, K. B., and Zaharko, D. S. (1970). Interspecies correlation of plasma concentration history of Methotrexate (NSC-740). *Cancer Chem. Rep.* Part 1, **54**:95–101.

Dicker, S. E., and Heller, H. (1935). Relationship between glomerular filtration rate and urine flow in the rabbit. *Science* **112**:340–341.

Dobyan, D. C., Levi, J., Jacobs, C., Kosek, J., and Weiner, M. W. (1980). Mechanisms of cis-platinum nephrotoxicity II morphologic observations. *J. Pharmacol. Exp. Ther.* **213**:551–556.

Elgee, N. J., and Williams, R. H. (1955). Specificity of insulin degradation reaction. *Diabetes* **4**:87–91.

Freireich, E. J., Gehan, E. A., Rall, D. P., Schmidt, L. H., and Skipper, H. E. (1966). Quantitative comparison of toxicity of anticaner agents in mouse, rat, hamster, dog, monkey and man. *Cancer Chem. Rep.* **50**:219–244.

Frenk, S., Antonowicz, I., Craig, J. M., and Metcoff, J. (1955). Experimental nephrotic syndrome induced in rats by aminonucleoside renal lesions and body electrolyte composition. *Proc. Soc. Exp. Biol. Med.* **89**:424–427.

Giacomini, K. M., Roberts, S. M., and Levy, G. (1981). Evaluation of methods for producing renal dysfunction in rats. *J. Pharmaceut. Sci.* **70**:117–121.

Hodson, C. J., Maling, T. M. J., McManamon, P. J., and Lewis, M. G. (1975). The pathogenesis of reflux nephropathy (chronic atrophic pyelonephritis). *Br. J. Radiol. (Suppl.)* **13**:1–26.

Hoyer, J. R., Maure, S. M., and Michael, A. F. (1975). Unilateral renal disease in the rat. I. Clinical morphologic and glomerular mesangial junctural fractures of the experimental model produced by renal perfusion with aminonucleoside. *J. Lab. Clin. Med.* **85**:756–768.

Jones, S. E., Lilien, O. M., and Rogers, L. (1963). Renal pelvic pressure and its relation to renal nemodynamics. *J. Urol.* **90**:357–360.

Kenney, J. M., Marinelli, L. D., and Woodard, H. Q. (1941). Tracer studies with radioactive phosphorus in malignant neoplastic disease. *Radiology* **37**:683–690.

Kirschner, A. S., Ice, R. D., and Beierwaltes, W. H. (1975). The Author's Reply. *J. Nucl. Med.* **16**:248–249.

Korobkin, M., and Shanser, J. D. (1974). Controlled unilateral renal artery hypotension in dogs. *Radiology* **111**:457–460.

Lannigan, R. (1963). The production of chronic renal disease in rats by a single intravenous injection of aminonucleoside of puromycin and the effect of low dosage hydrocortisone. *Br. J. Exp. Pathol.* **44**:326–333.

Lewis, E. J., and Couser, W. G. (1971). The immunologic basis of human renal disease. *Ped. Clin. N. Am.* **18**:467–507.

Long, R. G., McAfee, J. G., and Winkelman, J. (1963). Evaluation of radioactive compounds for the external detection of cerebral tumors. *Cancer Res.* **23**:98–108.

McAfee, J. G., and Subramanian, G. (1981). Interpretation of Interspecies Differences in the Biodistribution of Radioactive Agents. Third International Radiopharmaceutical Dosimetry Symposium, Oak Ridge, Tennessee. Oct. 7–10, 1980. Watson, E. E., Schlafke-Stelson, A. T., Coffey, J. L., and Cloutier, R. J., eds. Rockville, MD: Bureau of Radiological Health, 292–306.

McAfee, J. G., Grossman, Z. D., Gagne, G., Zens, A. L., Subramanian, G., Thomas, F. D., Fernandes, P., and Roskopf, M. (1981). Comparison of renal extraction efficiency of radioactive agents in the normal dog. *J. Nucl. Med.* **22**:333–338.

McGiff, J. G., and Quilley, C. P. (1981). The rat with spontaneous genetic hypertension is not a suitable model of human essential hypertension. *Circ. Res.* **48:**455–464.

McMahon, T. (1973). Size and shape in biology. *Science* **179:**1201–1204.

Mellet, L. B. (1969). Comparative drug metabolism. In *Progress in Drug Research,* Vol. 13, Jucker, E., ed. Basel: Birkhauser Verlag, pp. 138–169.

Miller, T. E., and Robinson, K. B. (1973). Experimental pyelonephritis: A new method for inducing pyelonephritis in the rat. *J. Infect. Dis.* **127:**307–310.

Oakamoto, K., and Aoki, K. (1963). Development of a strain of spontaneously hypertensive rats. *Jap. Circ. J.* **27:**282–293.

Oldendorf, W. H. (1974). Expression of tissue isotope distribution. *J. Nucl. Med.* **15:**725–726.

Potokar, M., and Schmidt-Dunker, M. (1978). The inhibitory effect of new diphosphonic acids on aortic and kidney calcification in vivo. *Atherosclerosis* **30:**313–320.

Ransley, P. G., and Risdon, R. A. (1978). Reflux and renal scarring. *Br. J. Radiol.* (Suppl.) **14:**1–35.

Rennick, B. R., Kandel, A., and Peters, L. (1956). Inhibition of the renal tubular excretion of tetraethylammonium and *N'*-methylnicotinamide by basic cyanine dyes. *J. Pharmacol. Exp. Ther.* **118:**204–219.

Rosenberg, L. E., and Segal, S. (1964). Maleic acid-induced inhibition of amino acid transport in rat kidney. *Biochem. J.* **92:**345–352.

Ryan, G. B., and Karnovsky, M. J. (1975). An ultrastructural study of the mechanisms of proteinuria in aminonucleoside nephrosis. *Kidney Intern.* **8:**219–232.

Salant, D. J., Darby, C., and Couser, W. G. (1980). Experimental membranous glomerulonephritis in rats: Quantitative studies of glomerular immune deposit formation in isolated glomeruli and whole animals. *J. Clin. Inv.* **66:**71–81.

Seiler, M. W., Rennke, H. G., Venkatachalam, M. A., and Cotran, R. S. (1977). Pathogenesis of polycation-induced alterations ("fusion") of glomerular epithelium. *Lab. Invest.* **36:**48–61.

Sharon, Z., Schwartz, M. M., and Lewis, E. J. (1977). The glomerular localization and transport of aggregated protamine-heparin complexes. *Lab. Invest.* **37:**43–52.

Sherwood, T., Swales, J. D., and Tange, J. D. (1971). Experimental renal papillary necrosis: Progressive changes on intravenous urography. *Invest. Radiol.* **6:**239–244.

Spector, W. S., ed. (1956). *Handbook of Biological Data,* Philadelphia: W. B. Saunders.

Stara, J. F., Nelson, N. S., Della Rose, R. J., and Bustad, L. K. (1971). Comparative metabolism of radionuclides in mammals: A review. *Health Phys.* **20:**113–137.

Williams, P. D., Holohan, P. D., and Ross, C. R. (1981). Gentamicin nephrotoxicity. I. Acute biochemical correlates in rats. *Toxicol. Appl. Pharmacol.* **61:**234–242.

Woodard, H. Q., Biggler, R. E., Freed, B., and Russ, G. (1975). Expression of tissue isotope distribution. *J. Nucl. Med.* **16:**958–959.

Index

MIX
Papier aus verantwortungsvollen Quellen
Paper from responsible sources
FSC® C105338

FSC
www.fsc.org

If you have any concerns about our products,
you can contact us on
ProductSafety@springernature.com

In case Publisher is established outside the EU,
the EU authorized representative is:
Springer Nature Customer Service Center GmbH
Europaplatz 3, 69115 Heidelberg, Germany

Printed by Libri Plureos GmbH
in Hamburg, Germany